A HISTORY OF HORSE BREEDING

BY THE SAME AUTHOR

Horses of the World
Horses and Their World
British Native Ponies
Successful Show Jumping
Huntsman of a Golden Age
Silver Spring
The Flight of the East Prussian Horses (*originally published as* Die Pferde mit der Elchschaufel)
The Foals of Epona
The Seventh Continent
Know Your Pony
Zebras

Translations
The Noble Horse
Cavaletti
The Asiatic Wild Horse
The Long Way Home
The Horse Today—and Tomorrow?
The Language of the Horse

Edited
Horses
The Pocket Dictionary of Horseman's Terms
Horses and Jumping

A History of
HORSE BREEDING

Daphne Machin Goodall

ROBERT HALE · LONDON

First published in Great Britain 1977

ISBN 0 7091 5593 X

Robert Hale Limited
Clerkenwell House
Clerkenwell Green
London EC1

Photoset, printed and bound
in Great Britain by
REDWOOD BURN LTD.
Trowbridge & Esher

Contents

Illustrations

PICTURE CREDITS

Edition Plassard: 2; Verlag A. Egger: 3, 47; Daphne Machin Goodall: 4, 5, 8, 9, 15, 19, 20, 21, 23, 59, 74, 77, 78, 79, 80, 82, 83, 84, 85, 86, 89, 90, 92, 94, 97, 109, 110; The British Museum: 6, 7, 12, 17, 24, 26, 29, 30, 32, 33, 34, 35, 38, 40, 41, 42, 43, 44, 52, 54, 62, 72, 73; Professor Rick: 10; Museum für Vor- und Frühgeschichte, Berlin: 11; Caisse Nationale des Monuments Historiques, Paris: 13; Musée National Suisse: 14; Mrs Louise Firouez: 16; the late Señor Ruy d'Andrade: 18; Oriental Institute, Chicago University: 22; National Museum, Denmark: 31; Egyptian State Tourist Administration: 37; Oscar Savio: 39; Bildarchiv Foto, Marburg: 45, 53, 60; Staatliche Museum, Berlin: 46, 57, 69; Etruscan Museum, Rome: 47; Jean Roubier: 48; Niels Elswing: 49; Scottish National Museum of Antiquities: 50; Leicester Museums: 51; Lometsch Verlag: 55; Réunion des Musées Nationaux: 56; *Die Rassen des Pferdes* by Graf C. G. Wrangel (Schickhardt and Ebner): 58; Kunstindustrimuseet, Oslo: 61; Ministry of Works, Edinburgh: 63; Rijksmuseum, Amsterdam: 64, 65, 68; Novosti Press Agency: 67; Biblioteca Marciana, Venice: 71; Landesdenkmalamt, Westfalen-Lippe: 75; Woldemar Klein: 76; Commercial Studios, Ipswich: 81, 91, 101; Sovfoto, Moscow: 87; Elizabeth von Kleist: 88; Ackermann: 93; Nova Scotia Communications and Information Centre: 96; Ivan Sanderson: 98; Mrs Elizabeth Bell: 99; Rutland Gallery: 100; Forres Sporting Picture Gallery, London: 103; A. C. Cooper: 104, 108, 111; Dr William Jones: 105, 106, 107; Sport and General: 112

For
Betty Kathleen Machin Goodall
Sophie-Charlotte von Uslar-Gleichen
John Crossley Hartle
with my love and thanks
for so much help given in so many ways

D.M.G.

When man discovered the wild horse
He tamed it, frequently misused it, sometimes worshipped and
occasionally loved it . . .
But the horse, wisely, learnt man's language.
Behind the eyes that seek far horizons,
There is the desire to serve without reward,
To work without gain,
And by giving, to please.
Is this a lesser creature than man?

D.M.G.

Author's Note

The following abbreviations are used:

h.h. (hands high), the British measurement of horses up to the withers, in units of 4 inches.

Stm. (Stockmass), the Continental measurement in centimetres of horses up to the withers.

Introduction

So far as I am aware, no history of horse breeding has been attempted. Many horsemen of today are unaware of the influence of blood lines in the pedigrees of their horses and even more people, be they horse-lovers, riders, trainers or admirers, probably have only the faintest suspicion of the great past and the care with which the ancestors of our horses and ponies were bred. Dedicated breeders have, of course, studied modern genetics. They have studied their chosen breed whether Thoroughbred, Hanoverian, Anglo-Norman or Standardbred Trotter.

In my book *Horses of the World*, I attempted to give brief outlines as to how these modern breeds of horses and ponies came about and their interrelation one with the other across the continents. In this book, I shall attempt to show how breeding started some 5,000 years ago, and follow it through to the major breeds of horses today. This is a reversal of the process used for *Horses of the World.* Probably the reason why no such attempt has been made before is because a certain amount of conjecture must be resorted to. We can follow the course of history, the invasions of peoples from other countries, the subjections of states, the interrelations and intermarriages, the exchanges of valuable stallions between potentates of distant lands and the development of some breeds of horses. But it is often difficult to assess with a degree of certainty if a breed of horses ceased to exist, say 3,500 years ago, if they merged, and how they might have continued as 'new' breeds. As far as possible I shall try to use only the facts, as far as they are known today. My conjecture and suppositions will be clearly stated.

A mass of pictorial reference is available. Generally the ancient artists were fairly reliable in portraying the events of their times. The events were often possible only because there were horses, and the horses were depicted just as they were used. Therefore we are able to learn *where* they were used and *for what purposes* they were used. The next step is to discover *where* they were bred and *how* they were bred. It is not a difficult matter to trace and check backwards 500 years, but it is a difficult matter to begin at the beginning, as it were, because we are still rather doubtful as to just what period should be termed the beginning. Only a few archaeologists seem to have seen the horse's connection with peoples' history and culture. Professor Hančar is one of these. German and Russian writers have studied and written about breeds of horses at certain periods, and a certain amount of material in Russian has been translated into

German. I propose drawing on this material. Confusion has often arisen.

Since the time when groups or herds of horses began to come within the denomination of a breed, such breeds have been named after the district or locality (stud) in which they are bred. More rarely is the breed named after the type of horse which represents the breed, or after an individual or potent sire. Some breeds are throughout history named after tribes and here it is important to appreciate that tribe *x* and tribe *y* may be using horses of exactly the same origin. We have a typical example in the breed name Trakehner once famous in East Prussia (now in West Germany and other countries). The Polish name for the Trakehner is the Masurian—an East Prussian horse, now bred by the Poles in the district of Masuria. Both horses are identical in origin and pedigree.

Language is an essential part of man's evolution and of his spiritual progress, and, since he came into contact with the species *Equus*, he is concerned with no fewer than 119 languages in Europe and Asia alone, by my own reckoning and knowledge of those peoples who *historically* had connections with *Equus*. Doubtless there are many more, because I have omitted all the North and South Amerindian languages. We know full well that many of these languages covered the horse and horse-breeding. Some of the 119 languages are 'dead' but before they died out, stone was cut and tablets engraved inscribed with information about horses, their breeding and training. One of the earliest of these 'books' at present known, of which there are several extant copies, was produced *c.* 1400 B.C. by a horse-trainer named Kikkuli, described as an Indo-European of Mitanni. However, since the Sacae tribes apparently did not reach the western regions of modern India until *c.* 500 B.C., this is an error in nomenclature. It seems probable that Kikkuli may have been a member of one of the Saki or Sacae tribes, and therefore would come under the Greek generalization of 'Scythian'. Kikkuli was a trainer of the first degree, and he described in minutest detail the training of some 30,000 horses over a period of 170 days. The text has been painstakingly translated into German by Professor Kammenhuber (see Bibliography).

As information about horses was circulated at first in the now 'dead' languages, and later taken up by new generations of people in 'new' languages which sometimes succeeded the old ones, or which changed completely when a country or nation was overrun or its original inhabitants migrated for one reason or another, it is perfectly clear that many misrepresentations came about. It is at the same time amazing how much knowledge has been handed down.

The variation in translation or in spelling of the name of a breed of horses or even of the district where horses are bred must never be overlooked. Thus we have the district of Corduène near Lake Van, once in the centre of the Hittite kingdom and later overrun or absorbed into other kingdoms and finally becoming a part of modern Turkey. During the centuries Corduène changed to Gordeyne, whilst some 200 years ago horses of the Gordeen breed (as their British importers called them) were imported to Britain via Aleppo to help lay the foundations of the now world-famous Thoroughbred breed of horses (see Chapter 8).

Another word is *anse* which is Babylonian[1] for horse and in very early times meant 'mountain horse' or 'ass from across the mountains' i.e. onager; later the Chinese word for Parthia was *An-si*. We also have the mystifying expression 'heavenly horses' (see Chapter 3) relating to those bred in Ferghana. The Chinese called them *Han-Hüe*, translated from Chinese meaning 'sweating blood', said of a superior breed of horses, but, says Hirth, this is possibly a transcription of some foreign sound. To my ear *an se* could sound to Chinese emissaries as *(H)an-(H)üe*. *Anse* certainly had a special interpretation for the Central Asiatic peoples of the third and second millennia since the animals with which they were all most acquainted were oxen and onager. Thus the definition 'sweating blood' could be a complete misinterpretation. It is a fact that in spite of research into a considerable mass of material I have come across 'sweating blood' horses only in conjunction with the Chinese connection with such horses. Although there is a rash which causes horses to sweat blood it seems unlikely that the superior Ferghana horses suffered in this way. I have found no reference to them in transcriptions from other Asiatic languages although there are references to the superior horses of Ferghana which are dealt with later in the text.

My attention was drawn to the possibilities of misinterpretation when I read the following in a 1968 Spanish magazine. Harnessed to a perfectly ordinary four-wheeled dog-cart were a team of apparently pure Spanish horses. This turnout was captioned: "Un 'Dox-hut'." Thus the word 'dog-cart' to a Spanish ear becomes 'dox-hut'!

The Asiatic Wild Horse has come to be called in most European languages the Przevalsky horse, after Col. N. M. Przevalsky the explorer, but the name itself is spelt according to the language concerned, in German it is Przewalski. The Mongolians in whose country this wild horse may still be found call it by quite a different name. Thus 'Statur' or 'Dzurlikadu', also 'Take'. The Kirghiz and Tartars call it 'Kertag', and in the Turfan region the name is 'Takky' or 'Jarwat'.

The country of Bactria was famous under Darius for its horses, and the town of Wazirabad—the centre of the breeding area—is variously called Bakus, Balch, Balkh, etc., so that a non-expert could be forgiven for supposing that the 'Bakus' horse is not the same as the Bactrian horses.

As a further example and one with which we shall deal in detail in the text, the word 'Scythian' (or 'Massagetan') is a collective Greek term for all those collective tribes of Sacae or Saki who inhabited the area, many

[1] Akkad, Sumer and Elam were states of pre-Babylonia.

thousands of miles across, from the Tien-shan Mountains in the east to the Carpathians in the west, from the second millennium to the present time. Although 'Scythian' is no longer used, the Cossacks or Kosaki still inhabit the Don—Dnieper region.

The word 'Arab' again is a collective term for the nomads of the Arabian peninsula. Their language is Semitic based. Now it is a collective political term and includes several African and Asian states. We therefore see how contentious the word 'Arab' is, when it is applied to a single breed of horses.

'Aryan' means a scientific, language group; it is not racial or national. The term is used for the early Iranian or Indian language group. 'Indo-Germanen' is a term first used in 1813. 'Indo-European' is a collective language group comprising the languages of many peoples from the proto-Indo-Hittite of ancient times to the modern languages of Asia and Europe. Tatars or Tartars are peoples of mixed ethnic, linguistic and cultural origin, nowadays speaking Turki languages.

We therefore see that by referring, as many writers have done, to an 'Aryan' people or 'Scythian' horses, we have entirely misrepresented the meaning and sense of the words. In the same way the use of the word 'Arab' instead of 'Eastern' or 'Oriental' for a *breed* of horses is—or was originally—entirely misapplied. There were hundreds of strains or local types of Oriental horses. Only 100 years ago there are listed 100 breeds (see Abidin; and Beltz). Now, of course, the word 'Arab' or 'Arabian' does convey to the average horse-orientated American or European a certain type, usually the Siglavy, of once desert-bred, well-bred animal. So that one can say that the modern conception of an Arabian horse is of a specialized modern breed. But it is a complete misconstruction to say that *this* breed has influenced many other breeds of horses. It may, indeed, have influenced a few, and its ancestors may and did influence some, but under what *term* or *breed* are its Eastern or Oriental ancestors to be regarded?

Even that famous of all modern breeds, the English Thoroughbred racehorse, is known in France as the *Pur-sang* and in Germany as the *englische Vollblut* to name but two countries with their own definitions or literal translations of their interpretations of an English word, in which, in fact, the word 'blood' (*sang* and *Blut*), does not occur! In a thousand years' time, will horse-lovers, if there are any left, know that these three words in different languages all refer to one breed of horses?

The reader may thus appreciate the truly vast complexity of a book dealing with the history of horse breeding. We are all entitled to our opinions, but in defence I will only say that I have endeavoured not to express an opinion, unless stated as such, but to unravel from the mass of material some sort of coherent account which is woven into the involved tapestry depicting the historical evolution of horse breeding.

DAPHNE MACHIN GOODALL

1. Nomads and Early Domestication

Early man in his wanderings can best be described as the proto-nomad, the forerunner of the genuine nomad of historical times who raised horses, cattle and sheep and, herding them, migrated to summer and winter quarters. But there were, too, the early settled farmers who also domesticated horses and cattle—and in Mesopotamia the onager. Indeed the nomad–herdsman–breeder and farmer–breeder still exist, whilst the specialist breeder has appeared in more recent historical times. Let us journey back into the mists of time and with certain known facts before us, gain some knowledge of the very beginning of the 'horse industry'.

Probably we shall never learn at exactly what period man's acquisitive nature decided him to gain dominion over the animals that surrounded him. When man evolved into a carnivore he caught and killed only the smallest unsuspecting animals and his hunting methods were to rely upon surprise. In time, however, hunting weapons were developed. At first these were heavy rocks or stones, and sticks or stakes with a heavy club end. Then the ancient flint-workers developed arrow-shaped flints which could be slotted into wood, and stone axe heads that could be attached to handles with thongs. Larger animals could then be hunted, and their bones could be used for weapons, they were models for the first forms of primitive art. There is no way of discovering when or where the first pit trap was dug in woodland in the path which the hunted animal was known to use. But there is little doubt that not only deer, sheep, goats and wild boar fell into these pits, but also horses and mammoths as well. Horses were driven over the cliffs at Solutré in France, where they left their bones for posterity to find. Whether they were really 'driven over the top' by hungry men, or fled in panic before packs of wolves or even a forest fire, we may also never learn. This disaster may have been deliberate or it may have been accidental. Animals were also hunted with the bow and arrow and the *boleadoras,* which is a thong with a weight at one end. It is thrown to entangle the animal's legs and thus to bring it down.

Probably at a very early time, man did have areas which he and his family group or tribe regarded as their own. This territorial claim is fairly common in most animals and birds. The great ice ages, especially the last glacial period *c.* 10,000 B.C., drove men and animals southwards and their descendants returned to their former haunts only when the climate allowed them to do so. Thus certain species were dispersed and forcibly distributed over wider areas. In some places the melting ice will have caused great floods of water, although an ice-melt like the previous glaciations took many thousands of years to materialize fully—maybe even as long as our present knowledge of history and ancient civilizations.

At some point in prehistory it must have dawned on man to exercise some form of dominion over the lesser animals. Perhaps he began by selecting herds of sheep and goats that were to be found in his immediate neighbourhood. The youngest members and perhaps some of the women would be organized to keep these herds in sight. Thus the hunter had not so far to travel in search of food. When it came to herding bison, the wild ox and horses, one can assume that this was men's work and the strongest and fleetest of the young men would be needed. No doubt only a sufficient number of animals were herded to provide for the immediate larder requirements of 'meat on the hoof'. Even this primitive form of domestication resulted in grazing in the neighbourhood becoming scarce. So the group or tribe would have to move on to fresh grazing grounds—and the proto-nomad herdsman was born.

It would be an oversimplification of a very complex development to assume that such evolution happened everywhere at the same time. The remains of very early prehistoric civilizations have been found in many parts of Europe, Asia and America and cultures are known in Europe and Asia dating from 600,000 to 10,000 B.C., the period of the last glaciation.

Clearly nomad herdsmen and farmers did not evolve at one particular time or place and many knew nothing of horses, being familiar instead with reindeer and onager. The later Stone, Bronze and Iron Ages did not happen all over Europe and Asia at one and the same time. Today Stone Age man is found in parts of New Guinea and Bronze Age man along the banks of the Orinoco in Venezuela.

To clarify the evolution of the nomad–herdsman–breeder I believe the following extract will be of interest. It concerns the methods and habits of the modern Laplander, whose life has not changed much in many thousands of years. I cannot do better than to quote from a book[1] written some 160 years ago. The writer mentions how the country of the Lapps is covered in parts with moss almost as white as snow, whilst another kind blackens all the trees:

> This moss, however, which deforms the country serves for its only support, as upon it alone the reindeer can subsist. The inhabitants who during the summer, lived among the mountains, drive down their herds in winter and

[1] *The Natural History of Quadrupeds*, vol. 2 (1811).

people the plains and woods below. Such of the Laplanders as inhabit the woods and the plains all the year round, live remote from each other, and having been used to solitude are melancholy, ignorant and helpless. They are also much poorer than the mountaineers; for, while one of these is found to possess a thousand reindeer at a time, none of the latter are known to rear a tenth part of that number. The reindeer makes the riches of this people. The female brings forth in the middle of May, and gives milk till about October. Every morning and evening during summer the herdsman returns to the cottage with his deer to be milked, where the women have kindled up a smoky fire, which effectually drives off the gnats and keeps the reindeer quiet while milking. The female gives about a pint, which, though thinner than that of a cow, is sweeter and more nourishing.

Upon the return of winter, when the gnats and flies are no longer to be feared, the Laplander descends to the lower plains; and as there are but few to dispute the possession of that desolate country, he has an extensive range to feed them in.

Their chief and almost their only food, is the white moss; which because these animals feed upon it is called *lichen rangiferinus*. It is of two kinds, the woolly lichen, which covers almost all the desert [tundra] parts of the country like snow whilst the other is black and covers the branches of the trees in great quantities. Dressed warmly in his deer-skin clothes, the Lapp drives his herds fearless and at ease, ignorant of any higher luxury that what their milk and smoke-dried flesh affords him. It sometimes happens, though but rarely, that the winter commences with rain and a frost ensuing, covers the whole country with a glazed crust of ice. Then, indeed, both the reindeer and the Laplander are undone. They have no provisions laid up in case of accident and the only resource is to cut down the large pine trees that are covered with moss, which gives them but little, so that a great part of the herd may perish without a possibility of assistance. It sometimes also happens that even this supply is wanting; for the Laplander often burns down his woods in order to improve and fertilize the soil which produces the moss upon which he feeds his cattle.

In this manner the pastoral life is still continued near the Pole; neither the coldness of the winter, nor the length of the nights; neither the wildness of the forests nor the vagrant disposition of the herd interrupt the even tenor of the Laplander's life. By night and day he attends to his herds. He gives himself no uneasiness to house his herds, or to provide winter subsistence for them; he is at trouble neither of manuring his grounds nor bringing in his harvests; he is not the hireling of another's luxury; all his labours are to obviate the necessities of his own situation. If, therefore, we compare the Laplander with others of a warmer climate, we need not pity him; he is blessed with liberty, plenty and ease. The reindeer alone supplies him with all the wants of life, and some of the conveniences; which serves to show how nature supplies necessities. So the poor little helpless native, who was originally, perhaps, driven by fear or famine into those inhospitable climates, would seem at first view, to be the most wretched of mankind—but it is far otherwise—he looks round among the few wild animals that his barren country can maintain and singles out one from among them, and that of a

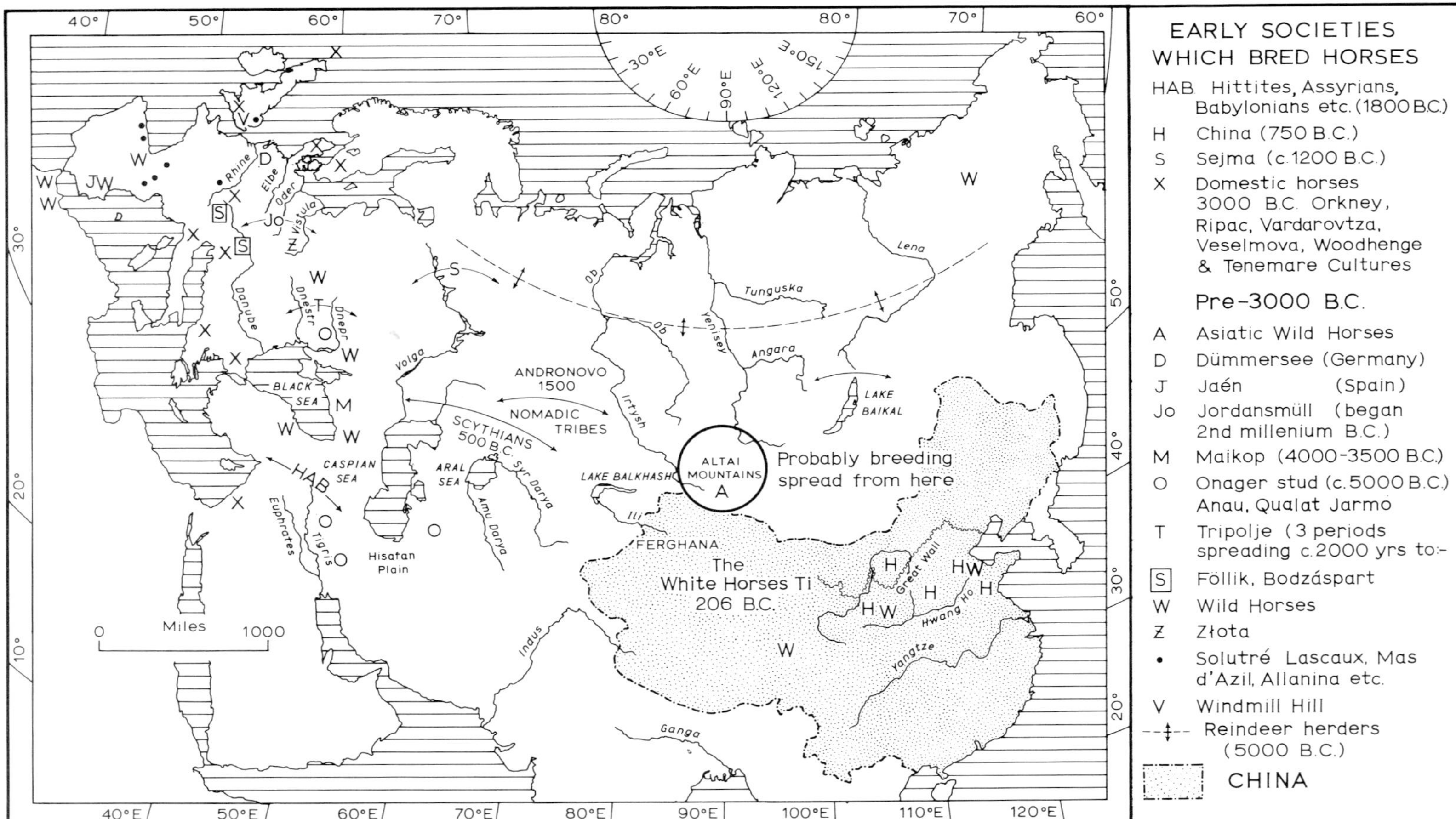

1. Map: Early societies which bred horses

> kind which the rest of mankind have not thought worth taking from a state of nature. This he cultivates, propagates and multiplies, and from this alone derives every comfort that can soften the severity of his situation.

Thus we can see how prehistoric man survived the Ice Age, how the nomad herdsman evolved and with him the semi-domesticated herds of different animals. Some herded horses long before others, and for some, inter-tribal relationships distributed knowledge and brought advancement. With this new knowledge and the fact that the herds had to be moved from one grazing place to another, there dawned in primitive man, besides the care for his family or group, the idea of caring for a species other than his own. So the wild ox with a weakling calf may have been brought to the cave or dwelling—or perhaps it was a foal that had lost its mother.

We continue with the quotation because it so exactly describes what must have happened many thousands of years ago, when the proto-nomad and subsequently the nomad herdsman began to herd horses.

> The reindeer of this country are of two kinds, the wild and the tame. The wild are larger and stronger [see next chapter], but more mischievous than the others. Their breed, however, is preferred to that of the tame; and the female of the latter is often sent into the woods, from whence she returns in-calf to one of the wild kind. These are fitter for drawing the sledge, to which they are early accustomed and yoked to it by a strap, which goes round the neck and comes down between the legs.
>
> The sledge is extremely light and shod at the bottom with the skin of a young deer, the hair turned to slide on the frozen snow. The person who sits on this guides the animal with a cord, fastened round the horns and encourages it to proceed with his voice, and drives it with a goad. In general they can go about thirty miles without halting, and this without any great or dangerous efforts.

The foregoing gives a very good description of the domestication of a primarily wild species, the taming of a draught animal and its subsequent service to the nomad herdsman. The reindeer is the first animal known to have been domesticated and that around 5000 B.C. The sledge described is also the earliest known form of transport. In Asia the onager and the ox were the next in line for domestication, whilst in the Mongolian Sajan Mountains, the reindeer, dog and later the horse followed. In some places—or cultures—the horse 'inherited' reindeer harness and in others, the cavesson of the onager or the yoke of the ox.

So we see how the nomad herdsman and breeder evolved and with him herds of the local variety of animals. Some herded horses long before others had even seen them. If the ancient Laplander could use reindeer to help him to survive, then early man could make use of his ox, onager, ass or horse. And probably the first domestication was as a pack animal, to transport the store of skins or dried meat whilst travelling from one grazing ground to another.

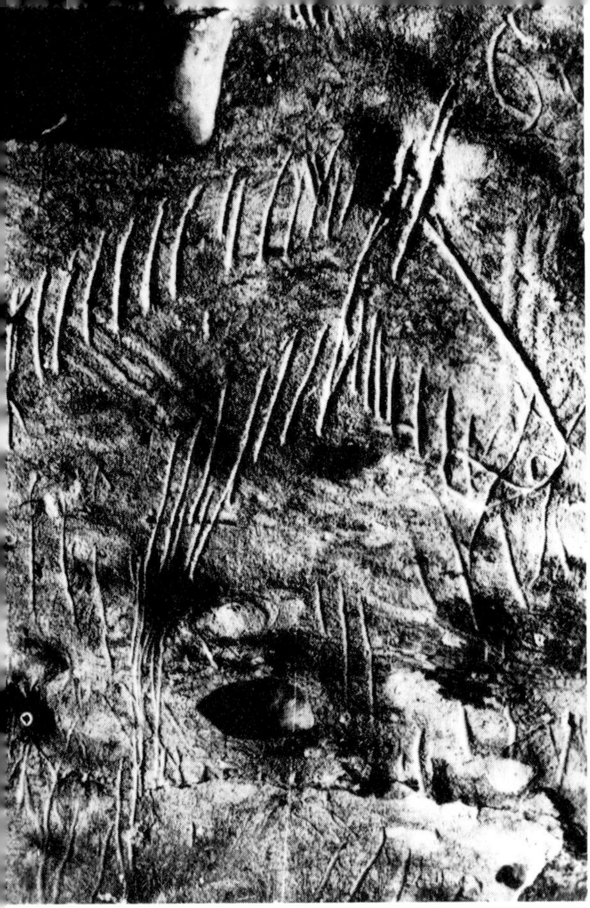

2. An engraving from the Grotte de Rouffignac, Dordogne, France, *c.*30,000–20,000 B.C.

3. A wild horse of *c.* 20,000–10,000 B.C. Cave painting from Niaux, southern France

Then something happened somewhere that caused the nomad to settle, at least for a while. Perhaps climatic changes brought good pasturage so that there was enough food for the herds all the year round. Lucerne and clover—alfalfa—were abundant in Asia, and elsewhere a primitive millet, rye and oats were growing. Thus some tribes took to a more settled way of life and the first farmers were born. Amongst the Cro-Magnon hunters there had been artists and we can see the result of their work in the caves of southern France and northern Spain. These early artists drew and painted with charcoal and ochre on the walls of their caves those animals which inhabited the neighbourhood. Very often they portrayed an in-foal mare or ox. And in two instances, at least, there are illustrations of the early horse-breeder, the horses being led. The horse-breeder is accompanied by his dog, just as he might be today. These illustrations come from rock engravings as far apart as Spain and China.

Most often these artists portrayed their own activities such as hunting, fishing, etc. It seems certain that most cave art had a semi-magical meaning and was connected with fertility rites, rites which were practised to ensure that the hunters were never short of game, and that the game never failed to reproduce. They even carved pregnant mares in ivory or modelled them in clay. When a people have a settled way of living and not everyone has to hunt or fish, to guard the flocks and herds, or to sow and reap the corn, the artist can emerge as the craftsman. He can fashion pots of all kinds from clay, he can decorate ceremonial bones—and later bronze—with symbolic designs, generally of animals and often of horses. The first 'architects' built megalithic grave chambers in which not only the dead were buried but also the horses that were to accompany the dead in the next world.

Britain was probably occupied by East Aurignacian peoples of the Solutré and Magdalenian periods—the Galley Hill people who preceded the Windmill Hill people of southern Britain, when this island was still joined to the Continent. Horses and people could then, around 10,000 B.C., walk across the valley of Dover from France and Holland, when the mouth of the Thames lay to the north of the Dogger Bank, which itself may have been settled.

The people of the Solutré and Magdalenian periods were superseded by Tardenoisian and other cultures, and by way of the more civilized cultures of the Tigris–Euphrates basin, northwards through Greece, farming methods spread slowly through Europe to Britain. And this is important because no one can breed animals successfully until and unless they have learnt to provide winter keep for their stock.

There were farmers in Asia and Africa before the inhabitants of the British Isles had discovered how to make pottery. It was around 8000 B.C. with the Natufien culture in Syria—the Natufien peoples were the ancestors of the Semites or Hamites—that the stone sickle was invented

4. Domesticated Tarpans on a Polish farm. The Tarpan is an ancestor of *Equus europaeus*, the all-purpose European pony

5. Wild horses of Europe—Tarpan *Equus przevalskii gmelini* Antonius in the Forest of Popiellen, Poland

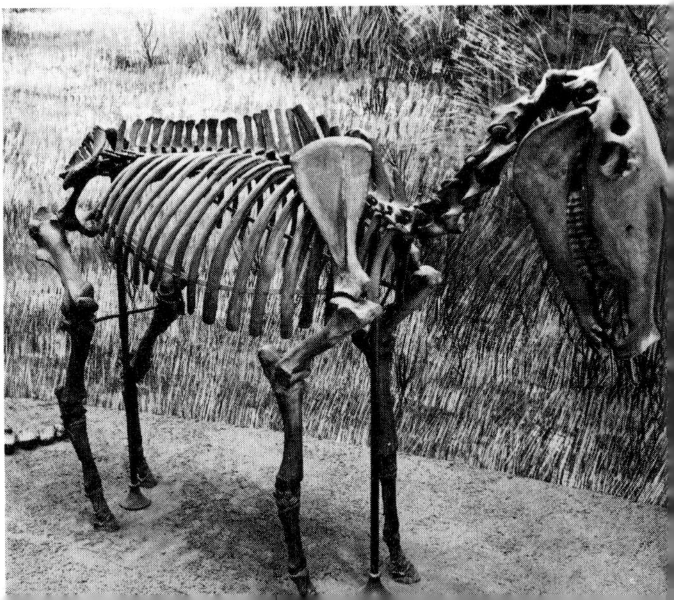

6. *Onohippidon*. Remains have recently been found which suggest that this horse was extant in South America at the time of the Spanish Conquest!

and the dog domesticated. Thus herds could be guarded and the wild corn varieties cut, harvested and stored, to be sown in prepared ground. Once this primitive farming became a recognized practice and the population more numerous, culture advanced at a fairly rapid speed. Man had taken thousands of years to reach this stage; the next stages were reached in a matter of only a few hundred years.

Just as agriculture involved migrations—fresh land had to be found for increasing populations—so a system of breeding and domesticating cattle and horses could evolve. It must have been difficult for these immigrants to adjust themselves to a rural economy. The damp Atlantic climate with wet summers and cold winters was quite different from conditions in Asia or around the Mediterranean. But the chalk downs of southern Britain provided the minimum of effort where would-be agriculturists could find areas of arable land and pasture.

To settle down and farm successfully meant a family and not a tribal arrangement. Hunters can and must move around in gangs, the nomad herdsmen probably had some form of co-existence with their fellows, but a plot of land will keep only a small group of people such as a family unit. Swine, cattle, sheep and goats could be herded by day and brought to a compound for safety at night, to protect them from wolves or marauders of any kind. On the moors and in the forests of Britain and Europe, the ancestors of the Exmoor pony, the *schweiken* of eastern Germany and Poland, the *konik*, etc., had found a habitat.

The early settler farmers of Britain, the Windmill Hill people of *c.* 2000 B.C., had a village culture, and probably the village itself consisting of simple houses of logs was built behind wooden barricades. This was done on the Continent. Roadways, too, were built of logs, laid side by side like railway sleepers. One such road apparently runs through north Germany, just south of the Dümmersee where some of the earliest European horse-breeders have left remains. In Britain the ancient highways are the Harroway which goes west to east across southern England, the Icknield Way going across East Anglia, and Peddars Way. There was no doubt a headman who was not only responsible for the safety of the village but he may have kept law and order too. Perhaps he directed the distribution of land outside the village and decided which beasts were to be slaughtered and which must be kept for breeding. No doubt his consent was sought when 'marriages' were arranged and cattle were exchanged to seal a bond.

Because today's society is masculine-dominated, one generally assumes that the head of a village would always be a man. The early societies and civilizations of Asia and Europe were not so sex-conscious nor so obsessed with women's inferiority. To them the sexes were far more equal and a matriarchal society was by no means uncommon. There are many examples.

The Windmill Hill people exchanged not only cattle but ideas with people on the Continent. They, too, were megalith builders.

With the arrival of migrations of land-hungry peasants of the Urnfield and Hallstatt cultures, some of whom had been displaced by the cold sub-Atlantic phase in central and north-eastern Europe, an agricultural revolution took place. The proper use of the plough helped to give well-balanced mixed farming and provided sustenance for a greatly increased population.

From Siberia to Norway, remains of the comb ceramic culture have been found, in Jutland, Silesia, Lower Rhine and eastern England where the Peterborough group advanced. These peoples needed space and this wanderlust agreed with their hunting–herding nomad life. In its latter stages, their culture was taken up with the progressive breeding of cattle and horses.

Around and after 2300 B.C., the Near East was in a constant state of flux. Both kingdoms of Mesopotamia and Egypt were destroyed by barbarian Asiatics from distant deserts and mountains. Towns in Anatolia were also destroyed and the ancient and civilized culture of Sumer was overrun. These barbarian invaders were mounted nomads.

Some centuries later, when the Marnians arrived in Britain, iron began to replace bronze. Their crafts and metallurgical and textile industries were conducted inside forts such as Maiden Castle and also in villages and farmsteads. Sickles, bill-hooks, axes and ploughshares were made by skilled blacksmiths. But in the more isolated areas, flint tools were still in use. These people bred sheep, goats and pigs; their cattle were larger with longer horns than the later Celtic shorthorn (*Bos longifrons*), but they were apparently smaller than the native wild ox (*Bos primigens*). Everywhere we find that the domesticated species was smaller than the wild counterpart. Perhaps crossing of the breeds was done consciously—just as the Lapp crossed the tame with the wild reindeer.

The reindeer culture had spread west and south, and the knowledge of herding and farming and of yoking animals to serve man's needs of traction power spread north and west. Pack loading must have come first. The earliest wooden plough developed from the hoe would have been pulled by man himself, harnessed to it by means of hide thongs. The transition from man draught power to animal draught power could have been rapid. Civilization, which had been slow to leave the Middle East, had arrived via southern Europe to Britain. From Britain and East Anglia it travelled to Belgium, north-west Germany, Scandinavia and Poland, then deep into the Ukraine and Russia.

There were now a number of races and although they must have spoken in many tongues and dialects, no written word has been found in Britain. In south-east Iran a team of archaeologists, led by Professor C. C. Lamberg-Karlovsky of Harvard's Peabody Museum, discovered the earliest known written tablets as recently as 1970. These tablets are said to be written in an early form of the Elamite language and reveal a civilization 1,000 years earlier than had previously been known. They are dated to *c.* 3500 B.C.

As these peoples grew in numbers, some more primitive, others more advanced, the influence of different cultures touched, overlapped and dissolved into new civilizations. The one used reindeer for riding and as a pack and draught animal; the other had first bred and used the onager or the ox and finally the horse. The one harvested rye, millet and oats; the other wheat, barley or alfalfa. The food they were given and the conditions under which they were herded and bred considerably influenced the evolution and development of all the domesticated animals. Skins, flints, tin, amber, ore, cattle and sheep could be exchanged. Where bride-thefts or intermarriages took place, here were the seeds for the beginning of a new culture. The one brought the knowledge how to sew with bone needles and the other an improved way to build houses or raise stock. Her gifts, or his gifts to her family, brought an improved breed of cattle to cross with the native breed and thus, as the seasons improved, the corn grew and was harvested, pasture was more carefully husbanded, and better domestic animals were produced.

Some peoples emerged from the mists with a genius for domesticating the onager, and later the horse. Known as Sacae or Saki they lived in that great horse-rearing country between the Oxus and the Jaxartes Rivers. They were, it seems, amongst the earliest horse-breeders of antiquity. From them descended the Indo-European family of peoples.

Quite a number of experts appear to think that the horse was first 'domesticated' as a religious symbol, but this is putting the event before the occasion, the 'cart before the horse'!

Of all the species, man has evolved into the most superstitious and credulous, but the nearer he was to nature, the closer was his affinity with the animals surrounding him and the more natural his native wisdom and association with those animals. Credulity and superstition belong to civilization—one might almost say to a corruption of society.

At the beginning, animals were kept in compounds for the simple convenience of having them at hand to shelter them from marauders. Obviously the occasions when neighbouring tribes met would be occasions for celebrations and the animals in the compound provided the wherewithal, whether they were sheep, cattle or horses. The early religions, or preferably mysticisms, involved the sun, that shining disc of light in the heavens which appeared with the spring and signified both birth and growth, and disappeared—at least sporadically—in the darkness of winter. So probably tribes met purposely to celebrate the appearance of the sun and Stonehenge and other similar monuments suggest this. Meeting places, first of wood, then of stone, were built in Britain and on the Continent. The best animals would be tended and kept in the compound for the celebration of spring. Therefore they would be bred in the compound and they would no longer be wild, although the horses were still very near to their wild forebears even though captivity caused a deterioration in size. Thus first a mystical, then a religious, significance arose with the celebrations of spring.

It must not be supposed that a primitive society is or was primarily concerned with mystic-magical or religious festivals. There was everyday existence to be coped with, and horses bred in captivity could serve other purposes than simply to provide a feast for symbolic rites. An unusual animal, for instance a white horse signifying light, could become a tribal totem, a sort of mascot. From the head of a tribe emanated light or wisdom, so the white horse would become a symbol of power—as indeed, in later generations, it did. If or when the light of that power was extinguished with the death of the leader, what would be more natural than to sacrifice the white horse? The edible parts would be eaten and other parts would be buried with the deceased head or king. A sacrificial platform has been discovered at Uppland, Sweden. The platform lay from north to south, and on it was found a horse's skull. The animal had been killed by a blow on the head, the lower part of two forelegs cut off at the knees and tail vertebrae were also found. "It seems," says Lundholm, "as if only the inedible parts of the animal were sacrificed." It also seems as if horses were not that common.

Undoubtedly these treasured animals in the compound would also serve as gifts—a horse could be used to placate a powerful neighbour. But in all instances it means that the horses in question were domesticated, they had to be haltered and led around, and one cannot do this with even a semi-wild horse. The time in history at which this was first done may be anybody's guess. My own opinion is that it was a great deal earlier than we think or for which we have, as yet, any real evidence. For such evidence cannot, in the nature of things, be preserved indefinitely. If the beautiful little ivory carving of an in-foal Tarpan mare, a jumping horse carved from reindeer horn, or the incredible combination of the etching of a wild horse on the skull of a real horse, all of which come to us from early Palaeolithic times, are anything to go by, then it is *not* impossible that already, remotely and seldom perhaps, horses were 'tamed'. However it is the presently accepted opinion that the horse was not domesticated in Palaeolithic times.

If indeed these people wore beautifully carved amulets of horses' heads, then it would seem that these animals meant something to them. Perhaps they admired their speed and agility, or perhaps they recognized and appreciated something of the character of the species that was to help man emerge into something higher than the animals.

Although we may choose to think of these ancient unknown societies as primitive or immature, yet it was a *venerable* primitiveness and not the decayed society upon which the horse is now about to bow his way out. Everything in the then known world had to be learnt and everything that the world had to offer had to be discovered—history had to be written, not excavated.

Thus the first steps were being taken in man's association with the horse, and their paths were leading in the same direction.

2. The First Strains of Horses

We have seen how horses could have first been domesticated and we must now try to discover what strains of horses were around in Asia and Europe. It is too early to think of breeds, but clearly from the fossil remains which have been discovered in many places, there were in existence different types of wild horses or ponies. In spite of changes caused by the glacial and inter-glacial periods, Europe, Asia and North America provided suitable conditions for the species to survive and subsequently evolve into the animal we know today. India, South Africa and South America appear not to have been able to support the true horse.

In India in the Siwalik Hills some of the oldest bones of fossilized *Equidae* have been found, whilst in South Africa there existed and developed three sub-species of the horse family; namely the zebra, the ass and the now extinct quagga, which like Chapman's zebra[1] appears to have possessed several horse-like features. South America provided a habitat not only apparently for the recently extinct proto-horse but also for two other remarkable *Equidae: Hippidium neogaeum* and *Onohippidium munizi*. An almost entire skeleton of the former is to be seen in the Natural History department of the British Museum. The skeleton stands 12¼ h.h. but the skull is very large by comparison to the body. There is too, a marked difference in the structure of the nasal region of *neogaeum* and *munizi*.

In a cave near Eggenburg, Austria, there have been discovered a great quantity of horses' bones, dated roughly to 50,000 years ago. One of the skulls indicated a horse of sturdy conformation whilst the teeth indicated that the animal belonged to the Middle Quaternary period. Similar animals are known from Remagen, Phöben, Nussdorf and Türmith. Otto Antonius gave to these horses the name *Equus caballus germanicus* Nehring.

The Eggenburg skull is described as possessing a markedly curved forehead and short muzzle, which differs from the above-mentioned skeletons but closely resembles the post-Ice Age Schussenried skull, though this was much smaller. The Russian Tarpan was about to evolve and Antonius asks if there is not a connection. Eggenburg = Schussenried = Tarpan—which would mean that the "Tarpan line had parted from the remaining horses during the Ice Age".

[1] *Equus burchelli.*

Differing from both the Eggenburg and Schussenried horse, the Russian Tarpan had a flat or deep forehead and Antonius notes that the same skull characteristic could be noted in the Near East onager, especially the Syrian–Mesopotamian and Iranian local breeds, and this characteristic distinguished them from their Far East relatives.

> Every Near East ass that I knew had this type of forehead but not a single Tibetan kiang . . . and this has nothing whatever to do with domestication or deterioration but was simply a characteristic of this wild horse form and its domestic descendants. And the characteristic short muzzle of the Tarpan, which is so often considered to be a sign of domestication—has in reality absolutely nothing to do with it. Quite apart from the fact that the Tarpan—like the Przevalsky horse—has this characteristic although it was in no way highly developed, it characterized many now extinct types of wild horses. Thus both the Eggenburg and Schussenried horses had even shorter muzzles than the Russian Tarpan. Furthermore, most of the North American quaternary horses were definitely short-nosed like *Equus occidentalis* and these were certainly not domesticated. And the skulls of zebras bred in captivity for several generations show no sign of any shortening of the muzzle.

Antonius quotes the instances of captive-bred wolves, foxes and wild boar where a shortened muzzle *is* a sign of domestication or deterioration. At this point, I would emphasize the mistake that is continually made, that is that captive-bred foxes are characterized by shortened muzzles and therefore a shortened muzzle in a wild horse means that it is feral and not truly wild; and because wild buffalo or caribou migrate, it follows that wild horses did so too. These kinds of statements are not only false and misleading but too often they have been passed on from one writer to another. Of course, to quote Antonius further, "It does not mean that shortening of the muzzle through domestication does not take place only that it cannot be proved and that the last remaining Tarpan herds were crossed with domestic horses cannot be argued, but undoubtedly a very strong wild blood component still remained."

Many illustrations portraying the Russian Tarpan are based more on imagination than fact. The never-ending dispute between European scientists, including also Russian, on the genuineness of the 'Forest Tarpan' must be mentioned. Because there are obvious differences in the shape of the skull, T. Vetulani distinguishes them quite rightly, I think, from the steppe Tarpan, although E. Skorkowski tried to prove that they are one and the same. Skorkowski and Schwarz declared that there never were forest horses,[2] and this, says Antonius, is easily refuted on the basis of palaeontological knowledge. These animals lived in forests, they were Tarpan in the zoological sense and can therefore be called Forest Tarpan. Right up to the end of the Middle Ages these Tarpan inhabited

[2] See Haiti horses, p. 225 and illustration 98.

forests in Europe, particularly in Reinhartswald and in the forests of East Prussia. They were obviously the ancestors of a number of native East European pony breeds. Naturally, both the Tarpan and the Przevalsky horse preferred the open steppe but as Antonius says, small herds were "sent into the forests"—or, rather, found their way for one reason or another into forest areas, which, after all covered and still cover a not inconsiderable area from Eastern Europe to the Pacific Ocean.

Horses were pushed out of the steppe by the invasions of man on their territory. They took to the desert if water was available, to the heath and to the forests—a dispersal and distribution forced upon them by man's encroachments. Pfizenmeyer wrote that just before the First World War, north of the forest zone, there were wild horses in north-east Siberia, and even in the tundra. Possibly some are to be found there even today—no one knows because no one has looked![3]

Antonius concludes that the Tarpan, including Vetulani's forest type, may be considered to be the basic ancestor of the Indo-European domesticated horses. The Tarpan type is easily recognizable in the oldest Bronze Age horses and those used by the Celts (described by Marek). It is recognizable in the ancient Germanic horse, described by Duerst, and the comparable Lower Rhine horse from the Dutch Terpen and the ancient Greek horse is also closely related. All these strains, says Antonius, give evidence that they are descendants of the Tarpan, even although they may have outcrossings of other blood in prehistoric times, "*improved selection on the one hand, and degeneration and neglect on the other, and the 'influence of local conditions' generally so that changes of conformation have taken place*". The italics are mine, for this is the ABC of the difference in breeds and will be considered later. He also says:

> It can be accepted that the horses of Turkish and Finnish–Ugric nomad horsemen who were closely connected with the Indo-European Scythians for centuries, came from the Pontian–Caspian steppes and were of Tarpan descent, although it is not possible to say what other blood they may have had.
>
> Amschler found skulls with wide foreheads and short muzzles as well as longer muzzles pointing to the Asiatic wild horse as ancestor. A similar discovery was made by Hanko, from skeletons in old Magyar graves, which belonged to the period of Magyar settlement. And the horse of the Avars belonged to the same group.

Wild horses generally have a larger brain cavity than domesticated horses. Though domestication can cause a reduction in the size of the brain, this is not the end of the matter. Lundholm[4] notes that early domestic horses had rather less brain *volume* than wild horses whereas horses

[3] See page 45.

[4] Bengt Lundholm: "*Abstammung u. Domestikation des Hauspferdes*" Zoo. Beitung. Uppsala 27.

of the Middle Ages showed the same brain volume as wild horses of similar size, and there are ponies, like the Veglia pony, which has an overall smaller skull and a larger brain than wild horses. These facts show that transformation in domestic horses is not uniform.

In nearly all species of animals and birds that have been domesticated, both muzzle and skull have grown shorter than those of wild forms. The muzzle is short because man prepares the animal's food for him, reducing the work of teeth and jaws which then grow small. The skull grows short because, by protecting the animals from natural enemies and hard weather, man has made unnecessary the full development of the animal's senses, and those parts of the brain in which messages from eyes and ears and nostrils are received lie fore and aft of the brain. Lack of use prevents the full development and thus the brain and the skull that encloses it are both shortened.

Regarding the outline of the profile from the fossil skulls so far found it is impossible to come to any conclusion as to a particular breed. The Przevalsky horse produced and still does produce both a straight and a concave profile line. Lundholm states that wild asses and domestic donkeys have the same or similar brain volume due to the *probability* that they have a monophyletic origin which makes conditions less complicated.

Early domesticated horses were smaller than their wild counterparts and this fact is shown most clearly from the teeth. One is forced to the conclusion that whilst wild horses had a free range from which to graze and could therefore develop to their maximum under the conditions of their habitat, domesticated horses whether captured as foals—after the mare had been killed—or bred in captivity were not so fortunate as to food. The agricultural knowledge and activities of the early farmers were of necessity very limited and it was a long time before a method of storing sufficient forage for the winter was invented and organized. One can well imagine that foals were stunted simply by lack of food—or of the right food. It has taken nearly 5,000 years to discover exactly what minerals and vitamins are needed by growing animals, in-foal mares, racehorses or beef cattle for example.

Summing up his 1955 thesis Nobis states that "only one wild horse form is proved to have existed during the Meso- and Neolithic periods and that is *Equus cab. przevalskii* whose autochone domestication in Central Europe is probable". He further says that a second form of wild horse, known as the 'Tarpan', cannot be verified. I find it difficult to be convinced by this statement of Dr Nobis, whom I met in 1957, when he personally gave me a table which does seem to be a contradiction. His work on the "origin and domestication of the domestic horse" based on fossil remains is exceedingly thorough, but when everything is taken into consideration only a limited number of remains have been found, one could scarcely expect otherwise, and that over a considerable period of time. These fossils are centred on north and central Europe including chiefly pre-war Germany. The fossil remains of *Equidae* in the British Isles, France, Spain, Italy, etc. have not been considered.

As I have already noted in my translation of *The Asiatic Wild Horse* (by Dr Erna Mohr) there is an almost obstinate resolution not to take into consideration the genuine existence of the Tarpan, nor to show interest in the work of Polish zoologists. The persistently expressed opinion is that the Tarpan is feral, a domestic breed gone wild and not a true wild horse, but no *proof* of the validity of this opinion is forthcoming.

If the Asiatic wild horse known as *Equus przevalskii przevalskii* Poliakoff—or *Equus cab. przevalskii* after Nobis—should become extinct in his native Mongolia and if specimens of this breed, bred in captivity in zoos, should be returned to the wild state in Mongolia and elsewhere, as zoologists state it is their intention, are we to assume that they are no longer wild but feral? If the Tarpan was wild in prehistory obviously some subsequent descendants will have been domesticated, but this does not prevent original wild herds from continuing to exist and to evolve separately from the evolution of domestic forms of their descendants. These separate evolutions can, and indeed must, be assumed during the course of the past seven millennia and perhaps even earlier.

The earliest large wild horse so far discovered was the Mosbach horse of Europe (*Equus mosbachensis*) which dates to about 100,000 years ago in the Early Quaternary. Several complete skeletons and skulls have been found. The cannon bones and some teeth of another, *Equus abeli*, were found at Heiligenstadt (former Province of East Prussia), but Antonius says that these big horses were replaced by the Middle Quaternary by the considerably smaller Eggenburg, Steinheimer, etc., horses.

Once again, I must draw attention to the fact that because fossil bones have not yet been discovered, one cannot state with certainty that big horses did not exist into a later period.

From La Combarelles, Dordogne, there is the little-known—or rarely produced—engraving of a horse's head. This is the head of a heavy Roman-nosed horse with what would appear to be a halter. The date is somewhat uncertain but it seems possible that this engraving was done during the period of Magdalenian civilization which was roughly 10,000 years ago. If this was a domestic horse then it shows a domestication several thousand years earlier than the Mesolithic painting in which no less than six horses are being led, from Canforos de Peñarubia, Spain.

Attention can be drawn to the magnificent amber horse which is dated to the end of the third millennium B.C. It was found near Friedeberg, Neumark, on the eastern side of the Oder. In this amber horse we have not only a cold-blood type of horse but also one with a distinctly concave face. It seems, therefore, possible that a heavier, but not necessarily *heavy*, type of wild horse did exist in north-eastern Europe and this is important to bear in mind when we come to the origins of cold-blood horses which were the ancestors of the much later war-horse and probably, too, of the universally distributed cob.

There is the much disputed question of whether horses were first ridden or driven. The answer, taking in Europe and Asia—both Near and Far East—as a whole, would seem to be that neither of these arts or means of transport can claim a 'first'.

As we have already seen, the earliest domestication of the horse was for its flesh and its milk; this was the sequel to prehistoric hunting. The very early reindeer hunters who domesticated this animal *c.* 5000 B.C. trained it to pull a primitive sled and to be ridden. The reindeer people inhabited forest tundra from Finland to Siberia and in the course of their seasonal migrations they must have come into contact with the steppe–tundra peoples who herded horses. After the last Ice Age most of Europe was forest–tundra–steppe, and Eastern Europeans as well had contact with the reindeer people. If a mare is sufficiently domesticated to be milked, she can be easily tamed to carry a pack, and if her back is strong enough she can carry a man. One has to remember that there were no roads, only the most primitive tracks and forest paths existed then. There was a very ancient track called the Harroway which wound right across the southern half of Britain leading from the lead mines of Charterhouse in the west to Dover, and this Way continued from Le Châtelet in France down to the Mediterranean. The Icknield Way is another. This Way cuts across Britain again from the west continuing through Dunstable and Icklingham to the east, with probably a branch to Dunwich across what was then land to the Dogger Bank (where trees have been discovered under water and upon which yachts are grounded even today) and so to Europe. Somewhere these tracks joined the one which runs from west to east close to the Dümmersee in north Germany and crossed the 'amber trails' from the west coast of Denmark where early man took North Sea amber down to the Adriatic. Another came from the Baltic Sea and crossed the Vistula to Biskupin, continuing down to an Adriatic port or even to the Black Sea (where amber from the north has been discovered). Flint implements have also been found in southern Europe and it is not entirely credible that man carried these on his own back.

The Wander Way goes through Tuva, north of Mongolia, from east to west. In Tuva there were reindeer, wild horses, possibly the onager, wild yak and wild camel and very possibly the earliest cattle breeders originated in this part. There is a further ancient way from Ephesus to Susa, and yet another, the Chinese Silk Road, used extensively by caravans in the Middle Ages but undoubtedly in use in much earlier times. It runs from the River Ili more than 2,000 miles to Ansi and Yumen, then it probably followed the Wei-Ho another 3,000 miles to the Yellow Sea.

Within fauna groups there were variations caused by the effect of climate and this applies at all times and at all periods down to the present time but particularly *then* to those areas covered by ice when much of the fauna became extinct or was isolated, when a local type evolved. Then as the ice retreated the area would be reinhabited and eastern and western fauna groups would meet. In this manner horse populations would mix and crossbred forms resulted whilst the older form may have disappeared. And in passing it is interesting to note that remains of some of these forms, particularly *Equus caballus plicidens*, have been found in the company of Neanderthal man!

Schwarz continues:

> *Equus cab.* Linaeus inhabited Europe into historical times; though not pure bred[5] it was found as the Russian Tarpan up to the nineteenth century and it is still found as a component in our domesticated horses. Osteologically it cannot be separated from the present Mongolian or Asiatic wild horse only the colouring is different.

The expression 'diluvial', now replaced by the Glacial or Pleistocene Period or Ice Age, is usually bestowed in geology to the latest division of time immediately preceding historic times and following upon the Neogene Period. This is roughly the period when we begin to learn something of the different strains of horses, and if we examine the material available, we shall also begin to have some ideas of the sort of horses which prehistoric man, the proto-nomad and eventually the nomad-herdsman and the settler could domesticate into breeding herds or droves.

During the last glacial period there was a great decline in wild-horse populations and large and small horses lived alongside one another for only a short time. Both Mesolithic and Neolithic horses were similar to the Przevalsky horse, e.g. about 13·3 h.h. (140 Stm.), so that possibly the Przevalsky was the first to be domesticated. That is, if we except the more likely domestication of the Tarpan. E. Schwarz has some definite ideas as to the sort of horses which our early ancestors used. He says[6] that European horses of the *Equus caballus* group differ in only one characteristic and this is their size. Also *Equus caballus* may be divided into three groups:

> A small late glacial to recent period type which can be regarded as the Western representative of the Mongolian wild horse [Przevalsky] having the average size of the latter; a medium type appearing everywhere during glacial and inter-glacial periods; and lastly the pre- and early glacial form of large

[5] The reader is already aware that Schwarz's opinion regarding the Tarpan differs from that of many other experts and, of course, the shape of the *head* is quite different to that of the Asiatic wild horse.

[6] *Jahrbuch des preuss. geo. Landesanstalt* 48, 1937.

7. Wild ponies on the Mildenhall salver, Romano-British period

8. *Equus europaeus*: Ewart's plateau horse exemplified in this Exmoor pony mare and foal

horses which were first discovered at Bawdsey in Suffolk and in the Auvergne and is known as the Mosbach horse.

Europe alone provided a vast area and there were undoubtedly several kinds of horses within the groups, as shown by Lundholm. Possibly they made up individual breeds, and droves of two types could graze in one locality as zebra do and without interbreeding. The area covered hundreds of thousands of square miles and the period lasted over many thousands of years. Horses lived amongst a variety of other fauna and indeed flora too, since this changed with the advance and eventual disappearance of the aftermath of glaciation. Clearly, within these thousands of generations the size of horses would vary. In Northern Europe they appear to have been relatively large. In Central Europe where remains have been fairly widely scattered, there seem to have been at least two types—one that was somewhat smaller than its northern neighbour which Nehring called *Equus germanicus*, and a pony type. Lydekker considers *Equus stenonis*, the southern-type horse, to be the early ancestor of both the Thoroughbred and the Trotter. Both of these latter breeds contain Barb and Arab blood and they are both oriental horses.

To summarize, the cause and effect of natural variations should be taken into consideration. The horses of the Ice-Age Mosbach period were large, as for example *Equus cab. robustus Pomel* 1858 (16·2 h.h. = 170 Stm.) and *Equus cab. ger.* Nehring 1888 which later deteriorated. These larger horses seem to have disappeared in Palaeolithic times. Large and small lived alongside for only a short time, and Mesolithic and Neolithic horses in Europe were similar to *Equus cab. przevalskii* Poliakoff 1881 (13·3 h.h. = 140 Stm.) which is found today only in Mongolia. Lundholm assumes that *Equus cab. przevalskii* was never domesticated in Europe and Runjancev says that there is no connection between *Equus przevalskii* and the Mongolian domestic horses. The slight material available favours the correctness of this view and agrees with finds of prehistoric horses that have been made in the region.

Bengt Lundholm in his excellent work[7] deals in the main with the origin of European horses. Stegmann (1924), who was one of the leading authorities of conditions in southern Russia, assumed that horses could be traced to five different wild types, which were domesticated at different places. The types given by him are:

1. The North European mountain horse *Equus europaeus*, which conforms to Ewart's plateau horse.

2. The European forest horse, *Equus robustus*. This type is West European and the heavier horse would be descended from it.

3. The Tarpan, *Equus gmelini*. The remnants of this type are the South Russian Tarpane. Stegmann says he heard in 1924 that this horse

[7] See Bibliography.

9. Tarpan yearling, photographed at Popiellen-Masuria, Poland

10. An ivory carving, dating from *c.* 30,000 B.C., of an in-foal mare, most probably modelled from the Tarpan, *Equus przevalskii gmelini* Antonius

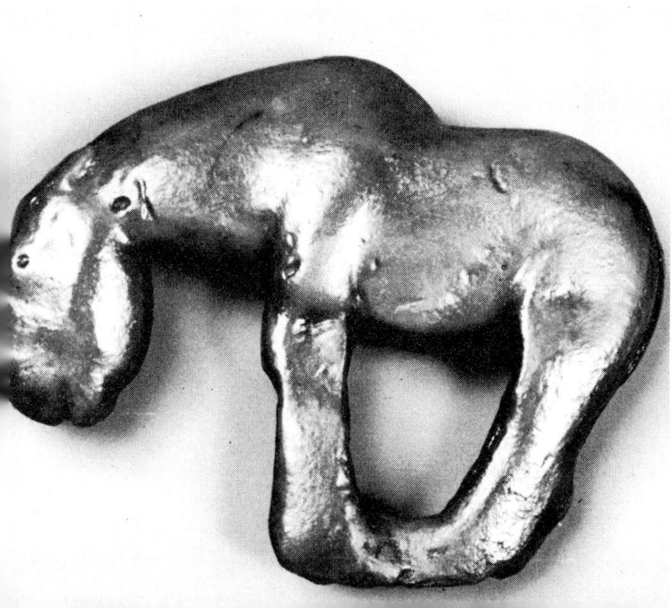

11. *Equus robustus*, probably the strain named *germanicus* by Nehring. Copy of the original amber wild horse from Woldenburg, Neumark, Germany

was still to be found in the wild state in the Southern Urals.[8] The East European domesticated horses are descended from these horses.

4. The Iranian mountain horse, *Equus orientalis*. This type conforms to Duerst's desert horse. The breeds of horses of the Near East developed from this strain.

5. The Mongolian steppe horse (or Asiatic wild horse), *Equus przevalskii przevalskii*. This type provided the material for Mongolian domesticated horses.[9]

There was, however, a sixth type which has often been overlooked.

6. *Equus stenonius*, today represented by the Sorraia pony, whose ancestors probably originated the native Spanish horse, and because of its distinctive convex profile may well have provided the ancestors for the Barb or Berber of North Africa as well. Some people consider that the Iranian mountain horse could be descended from the Tarpan and further, Stegmann's statement on the ancestry of the Mongolian domestic horse is in contradiction to that of Runjancev supported by Lundholm.

Nehring and von Reichenau produce a horse which they named *Equus germanicus*, whose fossil remains were found mainly, but not always, in the Rhine locality. *Equus robustus* and *Equus germanicus* appear to be similar types. Lundholm says:

> Certain Ice Age Swiss horses must be considered as belonging to this group. And in addition the Neolithic Swiss horses. So this type abounded during the main Glacial period (fossil discoveries in loess) and continued to live on until Neolithic times. The few available measurements for the horse remains found at Solutré in France, indicate that these horses too belonged to the *germanicus* type.
>
> In Europe, therefore, there were at least two types which belonged to the Western group. The others are Macalik (1912) who wrote about small Moravian horses.
>
> One horse also shows elements of the Eastern group, this is the Auvelinier horse. The tooth form reminds one of that of the Tarpan but the overall measurement is smaller. During the Bronze Age, there also appeared in Switzerland a small horse of Eastern origins.

Later Lundholm says:

> There are still many puzzling aspects which can only be resolved through renewed study, above all of East European and Asiatic finds of domestic horses in conjunction with the cultures with which they were found. It still remains to be ascertained whether it is a question of two centres of domestication, one in East Europe and a second in Central Asia, or whether wild horses were to be found in the whole area between the Russian steppe and Mongolia and

[8] Although it has been stated 'officially' that they became extinct during the last century.

[9] An opposite view to that of Runjancev which is discussed in the section on breeds.

12. *(right)* Horses and chariot from the Oxus treasure (*c.* fourth century B.C.) of Darius I of Persia. These are probably Tochar horses from studs in Hyrkanien and Choresem

13. *(middle)* A wild horse of *c.* 20,000 years ago—*Equus przevalskii*

14. *(above)* Two wild horses engraved on a reindeer horn from Schaffhausen, Switzerland (Magdalinien culture—early Stone Age—100,000–50,000 B.C.)

with them naturally both local forms and that these were domesticated. [E. Schwarz].

To sum up, there existed in parts of Europe and Asia the following strains of horses:

Equus przevalskii: the Asiatic wild horse, 12–14 h.h.
Equus gmelini: the Tarpan 13 h.h.
Equus europaeus: (so-called plateau horse, Ewart). 11–12 h.h. The Exmoor pony etc.
Equus robustus: Klepper, Døle, Cob 14.2 h.h. Welsh cob etc.
Equus orientalis: Tochar. The Arabian type. Caspian pony.
Equus stenonius; the horse of S. Spain and N. Africa. The Berber, Sorraia etc.
Equus europaeus and *Equus celticus* are in my opinion synonymous. Heights must be regarded as approximate. It could well be that before *Equus przevalskii* became extinct in Europe it had contributed to the evolution of *Equus robustus* just as *Equus gmelini* contributed to the evolution of *Equus europaeus, Equus orientalis* and perhaps, too, *Equus robustus*.

Before we turn our attention to the evolution of the horse in the hands of man, we should examine some of the material relating to the Far East and North East. There have been a number of expeditions into the deserts of Inner and Outer Mongolia, and one of these was led in 1928 by R. Chapman Andrews who wrote as follows:[10]

> We got skulls [at Baron Sog in Sumu] of a peculiar beast known as the *chalicothere* [see Simpson, *Horses*], a veritable paradox. It is a clawed-hoofed animal. The head and neck are those of a horse, the teeth like a rhinoceros, and the feet . . . instead of hoofs, the animal was armed with claws. Why such an anomaly was developed no one can tell. There must have been some good reason . . . but thus far the explanation is obscure.
>
> We found no trace of horses in the very old formations. This is a great surprise, for the unknown five-toed ancestor of the horses is one of the types which we confidently expected to discover. Four-toed horses are present in the Eocene of both America and Europe, and we are certain that the ancestral stock developed in Asia, but as yet it has eluded us. Nevertheless it must be there, although I have come to believe that it will be found in Southern Siberia rather than in Mongolia.

In 1923 at Ertemte, near Hallong Osso Dr Anderson[11] found fossils of *hipparion* in Pliocene or late Miocene horizons, the whole closely resembling the *hipparion* fauna of China proper. This country is now typical steppe. "If the Ertemte sand was deposited in an ancient lake, it means

[10] *Natural History of Central Asia.*
[11] *Essays on the Cenozoi of N. China*, 1923, Mun. Creo. Survey of China S.A. Mo 3.

Examples of *Equus orientalis: (top left)* 15. the Grosvenor Arabian (12 h.h.); (*top right*) 16. Caspian pony (10 h.h.)

17. *(bottom left)* Bronze statuette from early Gallo-Roman period. (Châlon-sur-Marne type found at Bourne, Lincolnshire.) Compare illustration 18

18. *(bottom right)* Sorraia pony (an example of *Equus stenonius*). This pony is from the unique herd belonging to the d'Andrade family. Note the resemblance to the Tarpan, and—a non-Tarpan characteristic—the concave face

that the climate of that period was immensely different from that of the present time, for this region is now characterized by treeless vegetation and small residual salt lakes."

In the fossil pits of Yen-Chung-Kou to the north is found the so-called *hipparion* fauna of the Pliocene, and this is characterized by an abundance of horses.

Mongolia is more than a million square miles in area and consists of a great plateau shaped like a basin. The Gobi is a rock desert with a 150-mile belt of grassland up to the base of the mountains where wild animals have their habitat and the Mongolian herds are grazed. The climate has changed many times although the country now enjoys a clear, cold winter followed by a hot summer. Professor H. F. Osborn notes: "For millions of years, this continent of Gobia was extremely favourable to the evolution of reptiles, mammals, insects and plants and probably birds." A remarkable feature is the almost complete absence of the *Equidae* although *hipparion* and *Equus* are found in great abundance near the southern edge of the Mongolian plateau.

Referring again to his 1928 expedition, R. Chapman Andrews says of the Artsa Bogdo area:

> On some basalt blocks we noted a striking display of prehistoric pictures. One group of these, found on a large flat fragment carries the following: two men, one with bow and arrow, several animals, one probably intended to represent a horse, another a reindeer, possibly a moose and fox. None of these animals except the fox and horse is to be seen within several hundred miles of this region at the present time, yet it must be that the artist was familiar with them. Either he belonged to a tribe which came from a region where these animals were native, or else the region itself was occupied in his time by such animals. It is entirely reasonable to believe the latter explanation and that the artist pictured animals living in his vicinity. This interpretation would mean that the people who made these drawings were not of the Mongol race and that the climate was other than it is now.[12]

The explorer E. W. Pfizenmeyer[13] made several journeys in search of mammoth skeletons into eastern and northern Siberia and he mentions fossil remains of wild horses and mammoth corpses on the Berezowka and also of the frozen carcass of a horse with very long white hairs which he heard about in Wercojansk. He writes: "There is hardly one place on the banks of rivers and lakes in the district of Yakutsk in which prehistoric remains of animals have been found that has not yielded skeletal fragments of the prehistoric wild horse", which he thinks is probably related to the Przevalsky horse.

[12] See also C. P. Berkey and F. K. Norris, *Geology of Mongolia*, 1927, Nat. His. of Cent. Asia., Amer. Mus. Nat. History.

[13] E. W. Pfizenmeyer, *Siberian Man and Mammoth*, trans. from German *Mamutleichen und Urwaldmenschen*, Leipzig, 1926.

19. *(above)* The Berber or Barb of North Africa (*Equus stenonius)*

20. *(top right)* *Equus hemionus*, onager, in Hagenbeck's Zoo, Hamburg

21. *(bottom right)* Asiatic wild horse *Equus przevalskii przevalskii* Poliakoff. A few still remain in parts of Mongolia

In the frozen earth of the bank of a lake in the tundra in the northern part of Yakutsk, the carcass of a horse was found and those parts which were sticking out of the ground showed a covering of very long greyish white hair. The district is the size of France but has only one inhabitant to ten square miles and the skulls of fossil horses are almost never properly examined—so that it can only be *assumed* that it was a prehistoric horse.

Earlier in the same book, Pfizenmeyer tells of the Lamut hunters from the Omolon River. "To our great astonishment Taitshin mentioned the wild horse. As zoologists thought wild horses existed only on the steppes of Central Asia, we received his statement very doubtfully, although Amuksan confirmed it by a quite professional imitation of horses neighing." The horses lived on the tundras bordering the forests of this vast area. In size it was equal to a Yakut horse and had long whitish-grey hair. "If the description were really that of wild horses, it was a puzzle how and when their tame ancestors could have reached this quite uninhabited Arctic region. And if they actually existed in the district between the two largest tributaries of the Kolyma, the Omolon and the Anjui—which had never yet been explored by any scientists [1913]." To date (1976), no expedition has looked for these wild horses, and one would imagine that they were not feral but truly wild. Amongst German scientists the idea that genuine wild horses may truly exist dies hard. It is just not acceptable to them.

Whilst we are still dealing with the earliest basis of breeding or origin of breeds of horses, it is necessary to remember that the onager, or wild ass as it is generally called in Asia, was amongst the earliest domesticated animals.[14] Hančar gives the earliest domestication of the onager, *Equus hemionus* onager, as *c.* 5000 B.C., remains having been found at Qualat Jarmo, and Antonius considers that the Standard of Sumer really shows the onager rather than the mule, as some have thought. The onager certainly seems the more probable since mules mean horses and thus crosses with asses or onager, and at the period of the third millennium B.C. we have very little proof that either horses or mules were to be found in Sumer or Elam. Meissner observes that during the earliest epoch both kings and temples had large herds of 'asses', "although a double pair ate 900 Sila [378 lbs] of corn per month, because there were no other draught animals". This allows roughly three pounds of corn per animal per day, so the animals were evidently stabled and not turned out to seek their own food. Antonius observes: "Perhaps the Accad-Semitic rulers of *c.* 2800 first brought the genuine African ass with them to this land when they replaced the Sumerian culture or at least for additional use. If this

[14] There are several varieties of onager and to lump them together as 'wild asses' is confusing, when there are genuine wild asses in Africa. The confusion arose from British army officers stationed in India.

22. *(above)* Tell Agrab—ancient Sumerian one-man chariot with four onagers. Driver is sitting astride on a saddle-like structure

23. *(right)* Asiatic wild horse at Whipsnade Zoo

24. *(above)* Royal Standard of Ur showing onagers, bottom row

supposition is correct, it is perhaps an indication as to the origin of the later Semitic rulers of the Near East."

The onager resembles the horse more nearly than the ass, and probably the Kültepe equines are onager as also the finds at Ur and Tel Halaf.

Thus we come to the puzzling clay tablet, known as seal 105 F, which was discovered by de Mecquenem during his excavations at Susa in 1930. This tablet is dated at about 2500 B.C. and is described with enthusiasm by W. Amschler (Director, Institute of Animal Research, Vienna 1935) as being "possibly the oldest pedigree in existence". Amschler appears to confuse onager representations with horses, and from this mistaken identity all subsequent assumptions are made. Much as it would help the compilation of this book, the entire theory must be refuted.

I quote from W. Amschler in *The Journal of Heredity* (1929):

> Some of the most important discoveries of this period were those of C. L. Woolley regarding the early civilization of the Chaldeans. Among the material brought to light in Woolley's excavations was a wonderful amber statuette of a horse [*sic*] which was found in the tomb of Queen Schub-Ad in Ur. This dated to early Sumeric times, about the fourth millennium B.C. Furthermore on the famous standard of the kings of Ur there are pictures of real horses drawing war carriages. This unquestionable evidence of existence of domesticated horses at least 5,000 years ago is among the earliest records we have. Furthermore it is clear that at this time the horse was by no means a new element in human economy. Woolley states that the houses of Ur are rounded off at the corners to facilitate the passage of the heavily loaded pack horses and donkeys in the narrow streets . . . Furthermore the horse had been present long enough to effect another architectural adaptation in the design of the steps of some of the houses which were built to make their use by riding horses possible. Pictures of hitched horses from Mari in the upper Euphrates valley show that Sumerian horses were used with a snaffle bit and not with a halter [cavesson] . . .
>
> In excavations of districts to the east of Ur—in Elam—discoveries of the greatest moment have been made.

Amschler then discusses the bone engraving of a rider on an *equine* calling it a horse. He says:

> De Mecquenem is also to be credited with discovery of another very important piece of evidence concerning the early history of the horse. This is a matter of especial interest to horse breeders and geneticists because it deals with the subject of horse breeding in old Elam and in the south-western highlands of Persia. The object in question is an engraved fragment showing a number of horses' heads arranged in a peculiar and very interesting manner. According to V. Scheil this seal in association with the as yet unexplained signs accompanying it, is an enumeration of horses. If, however, we examine this seal we discern the following significant facts:[15]
>
> 1. The horse-heads are situated in four horizontal rows.
>
> 2. The horses are of three categories as regards their manes: upright manes, pendant manes, and without manes.
>
> 3. If we examine the location of these three types we find that the three categories have a symmetrical arrangement with reference to the main diagonal . . . Note, in the fourth row, first head on the left, a sign which today is universally used in biology to indicate 'female'. In this latter case it is remarkable that the symbol is engraved upside down. In the same row the third head from the centre is accompanied by a sign which could easily be held to be a carriage with two horses. . . .
>
> 5. The heads with upright manes in my opinion represent wild stallions, those with pendant manes, tame animals and those without manes, mares.

[15] Even Amschler does not avoid the pitfall of passing off his opinion as 'facts'. See the Great Horse (Chapter 5).

6. It is held by Professor Scheil that the maneless heads are those of colts. That this is not the case is proved [*sic*] to my satisfaction by the fact that some of the maneless heads as the first in the top row and the second in the second row and the third in the third row are pictured of far too coarse and too powerful a build to be mistaken for colts. [The diagonal arrangement of heads may be also worthy of note.] On the other hand we discover in contrast some very delicate heads such as the fourth in the first row and the fourth in the third row are definitely maned. There are manes of different character, pendant and upright.

An analysis of the head forms is very significant to modern hippology. We can note the following facts:

a. A definitely ramlike profile of the head (head 1, row 1).

b. A ramlike profile only in the nose region (head 2, row 1; heads 2 and 3, row 3).

c. Perfectly straight profile (all heads of row 2).

d. Concave profile of Arabian [*sic*] type combined with graceful slender head (third and fourth head, row 1; fourth head, row 3; fourth head, row 4).

e. In addition to these definite differences of the profile line the size of the head in respect of the entire length and depth is interesting;

Long and deep head (head 2, row 2).

Medium long and deep (head 2, row 4).

Short and deep (head 2 and 3, row 4).

Long and slender (head 3, row 1). Connected with Arabian profile.

Medium long and slender (head 4, row 3).

All the above mentioned differences are regarded not as merely accidental differences of a mechanical or conventional nature but as studied reproductions of differences actually observed in the living animal. The head forms represented in this ancient seal are of types that today may be observed in the different races of horses. I need only note the existence in the Kladruber breed of a strongly ram profile and in the Arabian horse of a concave fine cut head to demonstrate that the differences recorded five thousand years ago in Elam exist still today. Intermediate shapes can also easily be found.

Furthermore these characteristics are familiar to the practical breeder and are known to be definitely hereditary. The dominant inheritance of some of these characteristics is so strong that they are recognized as 'prepotent' in inheritance as type of head has been observed to be and passed on for many generations. These head types are undoubtedly inherited as Mendelian characters in the horse in nature. Any one of these forms may become typical of a few animals, of families, or of a whole stable or artificial race through the agency of human selection. It is a matter of no little interest to breeders and geneticists that on this seal from Susa whose age must be fully fifty centuries, are shown all the profile types of horses existing today.

This ancient Elamic testimony regarding the horses existing at that time permits the conclusion that in Elam several original types and races of horses were then used. The horse heads are reproduced with wide, thick and plump, or with fine and aristocratic lines. There is no other alternative open than to classify the thick-set type with the Przewalsky horse (*Equus przewalsky* Poliakoff) and the slender form with the Tarpan (*Equus gmelini* Antonius). It is the considered opinion of the author that the early, heavy type of horse of western Europe is to be traced as a direct descendant of the Przewalsky type of wild horse. The heavy Persian horses are not to be considered as importations from Europe in prehistoric times but as the result of breeding native central Asiatic stock, which reached Iran with the Turanian invasions at an early time. The antiquity of this arrival of the horse cannot be determined but must have been very great. The more aristocratic type of horse was native to the highlands of Iran and the plains of northern Central Asia.

The early Sumerian and Elamitic civilizations, according to unequivocal testimony which is today available[16] knew and used the horse. It is the opinion of the author that the 'seal 105F' which has been discussed above is not as Scheil believes, an enumeration of horses but is a genealogical table of individual horses used for breeding in an old Elamitic stable . . .

Antonius says:

I cannot follow Amschler . . . besides different characters (or mainly) numerical signs of 19 half schematic *equus* heads. I would sooner have considered these to be hieroglyph, a simple sign symbol [mark], but it is quite possible that these symbols have the character of picture signs, thus that they possess the value 'horse'. But I can find absolutely no connection with Amschler's interpretation that it [the seal] represents a 'pedigree' of some royal stud. The inference that the upright heads with manes are stallions and those with hanging manes mares, and without manes foals, I have to refute on the ground of my knowledge of wild horses. The fuller and therefore heavier mane of the stallion even by upright maned species, is more inclined to fall than that of the thinner maned mare. But I would like to go even further and question whether these are horses . . . one comes to the conclusion that these are more like wild asses than horses. The wild ass [onager] possesses the most 'expressionless' mane possible—sometimes it inclines forward, sometimes backwards . . . it is thin and often only a continuation of the dorsal stripe . . . the thin neck too, is a characteristic of the Near East wild ass . . . and as long as no bones have been found I would not care to say that this refers to the horse. I do not wish to infer that the domestic horse which came via Iran to Mesopotamia could not have been present at this period but far-reaching theories on the age of horse breeding require to be based on stronger evidence. [*Zur Abstammung des Hauspferdes*]

Lundholm is far more emphatic: "Ein Beispiel für *völlig* grundlose Spekulationen in dieser Richtung ist der sog. Stammtafel der königlichen Pferde in Susa aus dem 4 Jt. welche Amschler (1936) gefunden zuhaben glaubt." (An example of absolutely baseless speculation in this direction is the so-called pedigree of royal horses in Susa of the fourth millennium which Amschler (1936) believes he has discovered.)

[16] In spite of intensive research such "unequivocal testimony" has not come to light.

Salonen on page 12 of his book *Hippologica Accadica* says that the sign for 'horse' is mane hanging (if not cut); the sign for donkey is upright mane, long ears; and the original sign for GIR_3, 'ass', is ass's head with neck and mane.

The evidence in favour of the onager and not the horse lies *exactly* in the indicated sparse mane and the thin neck but above all in the *characteristic* thick chin and jaw which seems to be present in all onagers (which I have seen and photographed at close quarters at Hagenbeck's Zoo, Hamburg) and which is seldom present in the horse, not even in the Kladruber!

The seal or tablet is Neolithic, about this there is no dispute. There are others in existence, namely those from Laguna de Janda, and by comparison this seal is very crude. People were capable of much greater expression and exactitude. When considering this seal and Amschler's interpretation we should not overlook the beautifully executed 'rein-ring'[17] made of electron. Without the slightest doubt the animal portrayed is an onager and the rein-ring was found at Ur, the capital of Sumer, the neighbouring state of Elam, at the same time as the seal, namely *c.* 2500 B.C. Only a valued domestic animal would be represented thus and it leaves very little doubt that the genuine wild ass or onager—and not horses—was in everyday use.

Amschler also refers to the Standard of Ur. He says: "Furthermore on the famous standard of kings of Ur, there are pictures of real horses drawing war carriages. This unquestionable evidence of existence of domesticated horses at least 5,000 years ago are amongst the earliest records we have."

I have studied the Ur Standard very thoroughly at the British Museum and by no stretch of the imagination—to one whose life has been spent with horses—are these animals meant to represent horses. Categorically they are not horses. In the first place, they are guided by nose rings as are the bulls or oxen on the same standard. The horse has a very sensitive muzzle and I would suppose him to go completely mad if a ring were attached to the soft part of his nose. Secondly the carriage animals are muzzled to prevent them from biting, which onagers are not shy of doing, and thirdly the tails are so obviously those of onager or ass. The head carriage, however, is that of the onager rather than the ass.

Hans Peter Lampe, who writes frequently for *Reiterrevue-International*, also makes the mistake of equine identity:

[17] There is some confusion of thought as to the exact use of this. The British Museum places it on the pole (of the chariot) and shows reins going through the rings. In practice no animal can be driven by this method. Either these are terret rings and would be situated on the yoke collar or on a pad (as in present-day harness) or, in my opinion, more probably it was affixed to the pole with thongs attached from the ring to collar, to hold up the pole. Similar rings are used today in pair harness.

> From Elam there comes to us one of the oldest records of the horse . . . Picture signs on a tablet record numbers of horses. Only horses' heads are to be seen. If the mane is upright, stallions are meant; if the mane hangs down mares are shown; if no hair is shown then a foal is meant. Near the horses' head left are finger-like pointers which have the value one, small circles are valued at ten. Since the tablet is read from right to left, the last row ends with "eight mares". On the reverse of the tablet the total of 185 horses is given.

I have never before seen any record of the supposed number of 'horses' on the seal. Let us examine the seal a little closer in conjunction with the Standard of Ur. First of all, part of the seal is missing and therefore some signs may be missing and I suggest that this was possibly a memo to the organizer of a projected movement, possibly a caravan. The writer knew quite well the meaning of his signs and wished to have a record of the order in which the animals were to go.

We decode the seal as shown opposite:

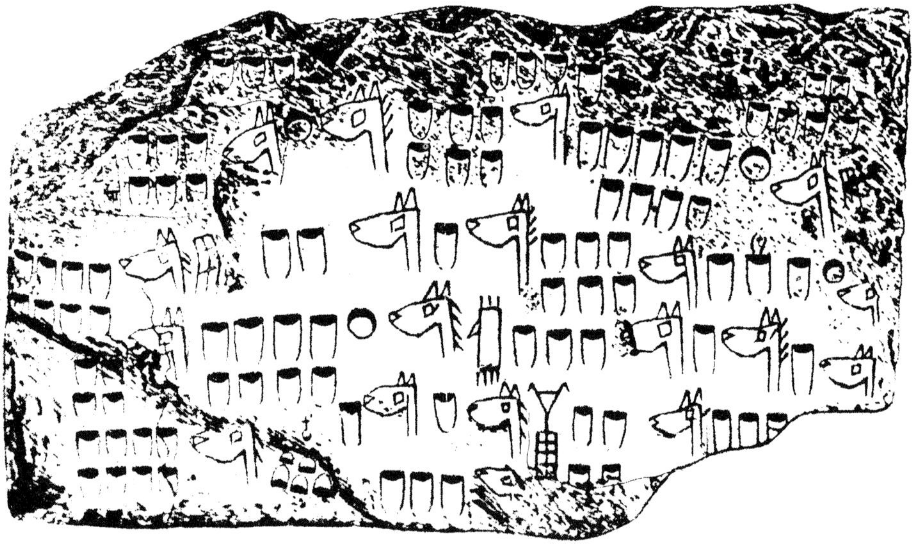

25. This engraved seal from an excavation near Ur in Mesopotamia is about 5,000 years old. A diagrammatic key appears opposite in which for simplicity the symbols have been arranged horizontally in sequence from left to right

The facing tables are a translation of the seal after the design of the Royal Standard of Sumer, 2500 B.C. Sumer and Elam were close neighbours. It seems credible that this seal was a memo to the organizer of a projected movement, possibly it represents the order of a royal Elamitic progression, perhaps to the king's summer palace. Possibly not all of the pairs were harnessed to chariots but may have been led. Obviously the originator of the hieroglyphic system knew quite well what his signs indi-

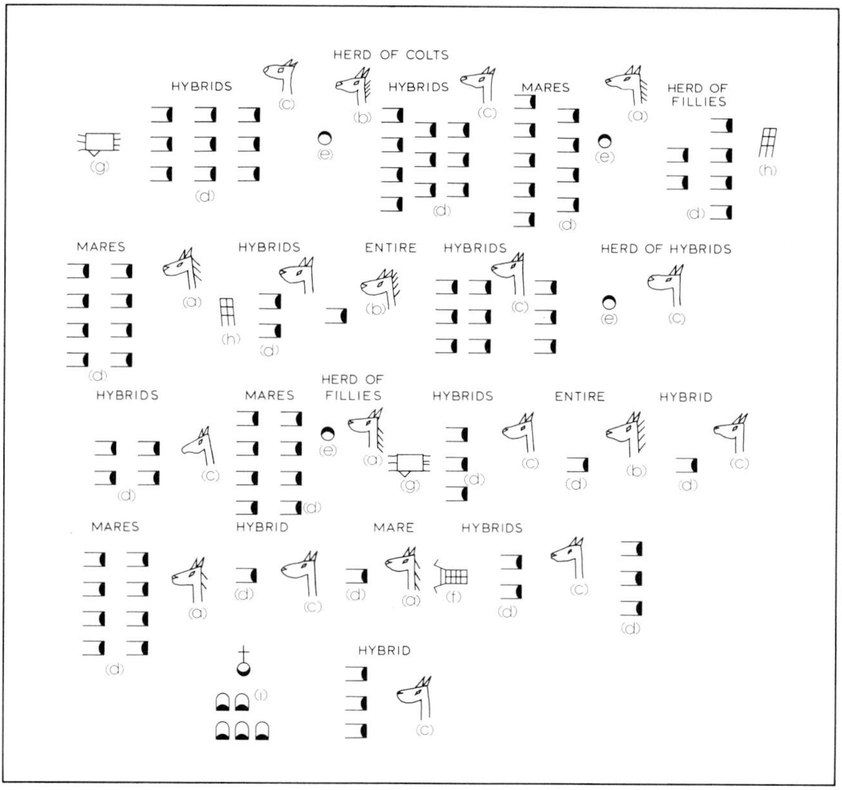

Key

It appears that the mares (a) are given only in pairs and sequences of 4, with one single pair and one pair of two + a herd of mares

Two single entires are given or pair entires (b) to a chariot and one entire in a herd or herd of colts

The hybrids (c) are given in pairs of three and two, and one single or two small pairs + a herd of hybrids.

Probably a chariot (d) or pair

Herd or tabun sign (e)

Waggon (f)

Possibly a terce waggon (g)

This (h) is known to be a Chinese sign for field. In the Kikkuli text (1000 years later) a field is a distance of 107 m.

Possibly an indication of amount of fodder required (i)

If the de-coding system is correct, it is noted that the entires are always separated by hybrids, as are also the mares.

cated. The equine representations are probably not horses but onagers although it is possible that *one* of the entires may have been a horse and the other an onager (lines 2 and 3). The horse was still very rare according to other sources. The colts would obviously be in the vanguard and the mares on their right are well protected. It seems possible that the two entires (lines 2 and 3) would be led, whilst the mare (fourth row) may have been single and not a pair and in-foal, and the two hybrids (third line right and fourth left from mare) may have been used to carry messages.

As to the supposed sign for field, the later Kikkuli text refers also to 'field' as a distance, thus: 1 field = 107 metres. Possibly this was the distance required to separate sections of the caravan.

The period is much too early to talk about 'pedigrees' which really date back roughly only 300 years, although horses as well as humans were 'selectively' bred much earlier. Even the vaunted pedigree of the Arab applied only to the mare, and the offspring of an Ubayan sire and a Saqlavi mare was a Saqlavi—whilst the son of a well-bred sire and a common mare was simply the 'son of a horse'. A well-bred mare was never mated with a worthless stallion. We shall return to this later.

But this is a mistake that often occurs—the application of modern rules, in this case pedigrees, to ancient customs. We even overlook that the geography of the place was other than it is today, that ancient cities, for example, were not then built in the middle of a desert.

The much written-of "horse of Anau" is probably more correctly described by the Russian A. Brauner who wrote: "Le cheval d'Anaou qui, à mon sens, n'est pas un cheval, mais simplement un âne sauvage de Syrie".

So to sum up (in the words of Antonius):

> In the Sumerian Old Mesopotamia, the wild ass [onager] has been proved as a beast of burden, in Turkestan—where bones have been found but no means of telling whether domesticated, although from the great quantity, indicate use as a beast of burden; at Elam on the shores of Mesopotamia and from the Susa seal the same thing can be said, and it is equally as possible in the Thebes mural of Asia Minor and Herodot's and Strabo's statements, so far as they do not specify the ass [donkey] that the onager is meant.

It does seem possible that when the domestication of the stronger and faster horse had taken place generally, plus the domestication and importation from Africa of the gentler and more patient donkey, then the use of the more intractable onager gradually ceased. This explains the latter part of Amschler's text: "Shortly before 3000 B.C. the horse disappeared from the lower Mesopotamia valley." It did not disappear, because it had never arrived.

"But," Antonius says, "remnants of onager domestication remained, above all for the breeding of mules." From Assyrian times there comes to us, a word *agalu* to which Christian[18] drew Antonius's attention. The

[18] V. Christian, *Archiv. für orient, Forsch.*, 10, 1934.

26. Hunting the onager. Assyrian relief, reign of Assurbanipal, 669–633 B.C.

former says it means a horse-like animal and a fast one. This last characteristic indicates the shy and speedy wild ass or his domesticated descendant. The *agalu* is different from the horse (*sisu*), ass (*himar*), mule and jenny (*paru, kudana*).

Christian has translated: "I am an agalu and I am harnessed beside a mule."

We are reminded of Assurbanipal's hunting scene from Kujundschik, where a number of onagers have been killed but one young colt is being led away on a neck halter—probably for use at stud. It was customary up to recent times when hunting the Asiatic wild horse to kill the dam and capture the foals (which were then exported to zoological gardens), but in earlier times this method may have been the foundation for domestication.

It is possible that during Roman and Greek times the faster onagers were still used to breed mules, and Plinius mentions that the cross from the onager and mare resulted in a hybrid which was marked by speed, hard feet, and a lean body with a mind of his own. There was mule breeding from the ass × mare cross, and it is not always easy to distinguish one from the other.

A population of Europoid type is the first known in southern Siberia. To the east and south, people with Mongoloid characteristics occupied the region. The origin of the first named is, however, not known and their Asiatic history is known only from the beginning of the Neolithic period "when they started to use metal for the manufacture of implements, to rear domestic animals and to cultivate useful plants, and thus set out on the process of development towards a productive economy".[19]

After some 2,000 years these people were followed by the early nomads with their progressive stock-rearing economy. "This economic progress," says Gryaznov, "was accompanied by a general rise in the level of culture and by a high standard of artistic production." During the early and middle Bronze Ages until about 1200 B.C. there was probably a comparatively peaceful way of living within the steppe countries. And, up to this time, those graves that have been opened by archaeologists show no sign of nomad horsemen. Further west and south, of course, civilization had advanced further and the "dynamic core was further South in modern Turkmene, where cattle-breeding tribes had concentrated", (K. Jettmar)

To the north of latitude 40 there was a great area of cultivated farming land which was to a large extent settled. This area stretched from the south of Russia to the "gates of Dsungaria". "These people, who belonged to a European type lived from rye and wheat growing and cattle breeding. Horses were simply a source of food—ribs of horses have been

[19] Mikhail Gryaznov, *The Ancient Civilization of South Siberia*, Barrie & Rockcliff, Cresset Press, London.

found in the graves," (K. Jettmar). Homer tells us that the mares were milked.

This Andronovo culture is the most comprehensive and uniform complex of the Bronze Age. Whilst the women concerned themselves with farming the cultivated fields, the young unmarried men herded the cattle on the steppe and pasture lands sometimes far distant and so discovered new lands to settle. Some form of exchange or trade took place between the different groups of herdsmen and apparently they had no serious enemies. Even older hunting cultures had retired to the mountains. From finds in the Gobi not completely destroyed by erosion it appears that there were also cattle breeders to the east of Mongolian type who were also a comparatively peaceful people, and further east a Chinese agrarian culture is suggested. Gradually the Andronovo peoples began to breed animals other than cattle, especially "sheep and horses which combined with farming required more people to help with the work", (K. Jettmar).

Towards the end of the Neolithic period, stock-rearing in the steppes became more of a domesticated business. It took many centuries of association with animals before man learned to obtain a maximum economic return and this included animals bred for milk, wool, meat, for transport and for ploughing. Here, in the steppes, this process of domestication was completed during the first half of the second millennium B.C. when, instead of following their forefathers as stock-rearing nomads, the inhabitants changed to a pastoral and agricultural economy.

So it was here that possibly a system of selective breeding was first practised and during the succeeding centuries different types of horses were produced. In the first instance the nomad stock-raisers may well have exchanged animals to improve the quality of their herds. In the matriarchal society which generally prevailed in Asia, the young men were accepted into the tribes of their brides, but with the political changes caused by territorial expansions when it became necessary to have a mounted warrior vanguard, the patriarchal system evolved. With this new system many more horses were needed.

It is generally thought that the early domesticated horses may have had rather weak backs and therefore they could not be ridden. Later on, with the expansion mentioned, the conformation of local types of horses was obviously strengthened, so that they were ridden and were used for forages and invasions into other territories.

With the change to pastoral domestication, herds of livestock had to be larger. Although summer pasturing in the steppe was usual, in winter the stock had to be yarded or housed. And, most important, fodder had to be grown for them, but with the still comparatively primitive methods of farming, productivity and quality would be low. All references indicate that the *domestic* animals were smaller than their wild counterparts (see page 26).

The Andronovo culture (which followed the Afanasyevskaya culture) is placed from the middle of the second millennium B.C. This culture covered the area from the River Ob to the Yenisey and in all essential features it is uniform. There must have been connections with the peoples from farther north, because the gold used on harness, etc., of a somewhat later date was probably found in the bed of the River Lena. In the remains of horses which have been found in burial mounds of a later period, it is fairly conclusive that different types or strains of horses began to appear in different areas. But let us hasten slowly! Now the pattern again changed, as for instance with the Karasuk tribes, who reverted to semi-nomadic stock-farming after the sedentary methods of agriculture of their ancestors, possibly because their home grazing grounds were exhausted by over-exploitation. From the quantity of bones of horses found in grave burials, it seems fairly certain that the Karasuk people owned a considerable stock of horses. These will have been the typical stocky steppe pony, not far removed from its wild ancestor the Asiatic wild horse and similar to the horses of the "Horsemen under a Tree" and the Tuva bone horse, although these examples are actually dated a thousand years later and will be referred to again.

Here, in the fifteenth and fourteenth centuries B.C., the horse began to play a greater part in the economy as a means of transport. It is thought that, in this region, it was *at first* used for draught because the weakness of the back could have made it unsuitable for riding. Gryaznov indicated that bone cheek-pieces—the earliest known—have been found in the steppes of Eastern Europe and Kazakstan which, he says, provides "evidence of the first attempts to use the horse for riding; and in settlements of the Karasuk period in the Yenisey and Ob valleys excavation has yielded bone and bronze cheek-pieces of an improved pattern with three holes, from bridles of early Scythian type, still without a metal bit". This seems to indicate once again that these were ponies since the 'bit' was most probably a leather thong, which, of course, would eventually rot and would therefore not be amongst the bridle parts in the grave. The discovery of these bone cheek-pieces over as wide an area as Eastern Europe and Kazakstan, and indeed in the Heathery Burn cave in County Durham in Britain where bits made of the antlers of the deer from the Early Bronze Age were found, indicates a definite, if remote, cultural connection.

From the Canforos de Peñarubia we see horses being led, probably only by a halter or headstall. M. Piette, the French archaeologist, is of the opinion that for ten or even twenty thousand years the halter was exclusively used—as it could well be on ponies. Today the halter is still used, as, for example with the American hackamore bitless bridles and the primitive rope or hair halter used by the natives in Indonesia.

We come to the late Bronze Age of southern Siberia. The people have changed to a completely nomadic herding life. "They were," says Gryaznov, "skilled horsemen. They had long been accustomed to using wheeled transport, such as carts drawn by a pair of oxen,[20] and the waggon [*araba*] was the only home they knew as they moved about the steppes."[21]

We may assume that now, around the tenth century B.C., breeding horses had become a serious undertaking. One cannot become a skilled horseman on an 11 h.h. steppe pony! So we are justified in thinking that selective breeding was practised and that bigger horses were being bred. The nomads moved around in hordes, again an indication that many horses were available and these were probably herded, just as they are today in *tabuns* guarded by mounted horsemen. As these nomads moved around, becoming more and more mobile, better grazing became available, and so the horses improved and were essential to the skilled horseman-turned-warrior.

[20] Of which there is an excellent photograph in his book.

[21] The Araba was a heavy hooded waggon used by the Tartars. In Arabic and Persian, *Arabah*. This may be where the name 'Arab' (for horses) originated, i.e. the *Arabah* horse, the horse accompanying and protecting the araba!

3. Domestication in China and Europe

The Far East and China merit a section on their own, not because the evolution and development of the horse had any particular influence on future breeding in Central Asia or Europe but because of the peculiar isolated position of this vast territory, and the large numbers of horses contained in it from earliest times.

There is archaeological evidence that horse and man were acquainted since prehistoric times. In fact, remains of a large Pleistocene horse, *Equus sanmeniensis*, were found amongst the bones and skeleton of *Sinanthropus pekinensis*.[1] The remains of *Equus przevalskii* was found with the tools of Palaeolithic Chinese man, and can probably be counted as the ancestor of future horses in China.

Horse bones were found, together with species of ostrich, elephant, rhinoceros, deer and goat, at Sjara Asso Gol on the southern borders of the Ordos region, about 50 metres below the present level of grassland. Palaeolithic hunters were responsible for these deposits. The Neolithic period in China continued later than in the West and overlapped the Early Bronze Age in Asia.[2]

There is a widespread belief amongst ethnologists and early historians that the earliest inhabitants of the Far East (China) learnt about horses from their central Asiatic neighbours. Sinologists, however, state that this belief is without foundation. The horse was native and wild in a number of places and was caught and tamed by Late Stone Age man in the same way as his contemporaries caught and domesticated horses in Central Asia and Europe, with but one difference—it was not ridden until more than a thousand years later. A few bones found on early Kansu sites seem to indicate that the horse was hunted rather than domesticated before 2000 B.C. But there is a representation of a horse on a clay vase also from the Neolithic period.

The late Neolithic Ch'eng-tse-ya culture of Shantung apparently indicates that the horse was domesticated. Creel[3] says that all Indo-Chinese languages have a single word for horse, and he thinks that the Neolithic inhabitants were well acquainted with wild horses which they caught

[1] Boule, Breuil, Licent and Teilhard, *Le Paléolithique de la Chine*, 1928.

[2] William Watson, *China*, Thames & Hudson, 1961.

[3] *Studies in Early Chinese Culture*, 1937.

and tamed. Erkes[4] is of the opinion that knowledge of horses was taken by the pre-Indo-Chinese to Central Asia and the Ural Altai districts, not the reverse, since the word for 'horse' in Mongolian, *morin*, and the Korean *mal* are based on the Chinese *mar*. The inventor or discoverer of horse-breeding was a mythical person called Ch'eng-ya which means 'capable to drive'—and this at a time when horses are thought to have been around only in the wild state! "If," says Erkes, "horse-breeding had been brought to China from outside, then there would have been no necessity to worship the discoverer." That the "inventor of horse-breeding" is, in fact, the embodiment of an historical memory is shown by the fact that in the *Lü-shi* Ch'un-ch-iu the "inventor of cattle-breeding" is also named. This also originated during the Neolithic period on Chinese soil.

So Ch'eng-ya seems to represent a memory dating to the prehistoric domestication of the horse. Then there is yet another Inventor of Horse-breeding whom Hsün-tse calls Ch'eng-tu who was a grandson of the Hsie—the mythical forebear of the Shang—thus making Ch'eng'tu a demi-god of the Shang civilization, which was a centre of horse domestication, the cultural kernel of which was accepted as being in east Honan. This discovery dates to the fifteenth year of the Hsia Emperor Hsiang, 1928 B.C. The oracle bones of the Shang period tell of hunting the horse.

There also seems to have been a second centre of horse-breeding whose inventor was honoured by the Chou, a semi-nomadic folk interested in farming and raising cattle. The Hsiao-jen, the stud-manager, and the head groom in summer, made sacrifices to "the first who bred horses; his identity is otherwise not known". Since the Chou had no name for him and he had a different personality to that of the Shang, it would seem that they got their ideas of horse-breeding from another culture. "Perhaps," says Erkes, "from the Hsia, which was equally an ancient centre of horse-breeding . . . Even the method of sacrificial rites confirms the view that horse-breeding on Chinese soil developed from the domestication of native wild horses."

The domestication of these native wild horses took place in several localities of ancient Chinese culture and amongst neighbouring tribes. Hsiang-Tu, the ancestor of the Shang, was the first inventor in east Honan. The ancient Hsia district in north-west Honan and south Shansi were also localities. Then Chin in the Hsia district is often mentioned in ancient literature for its wealth of horses, as well as the experts on horses who came from Chin and its neighbour Chao, both belonging as already mentioned to the Hsia culture.

And Hsia seems to have originated the invention of horse-drawn chariots. Even the sign 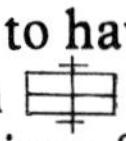 for chariot or cart dates back to prehistory, like similar representations from prehistoric Europe. But as in other places,

[4] Edouard Erkes, *Das Pferd im alten China*, 1940.

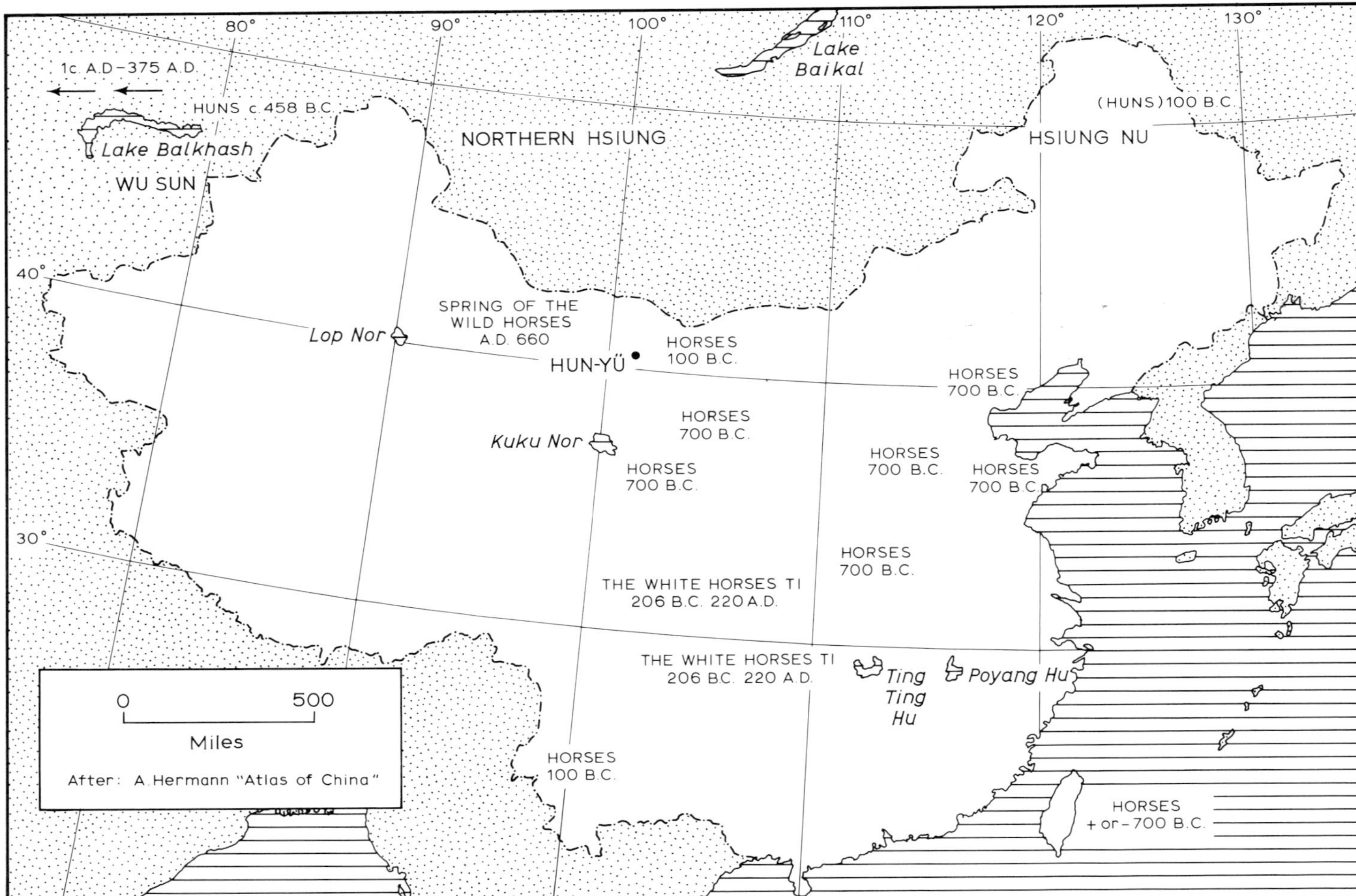

27. Map: Horse-breeding areas in ancient China

the harnessing of oxen to carts preceded horses, whilst a 'whip' by the name of Hän Ngai is credited as Inventor of Driving Chariots.

Around the beginning of the Hsia period, *c.* 2000 B.C., we learn of the colours of horses. The Hsia, it seems used (bred?) white horses with black heads, and their neighbours, the Chou, yellow (chestnut) horses with red manes. The same source mentions that in the Ch'in period it was the custom that the Emperor's war chariots were drawn on special occasions by *po-lo*, white horses with black manes, which, says Erkes, was a transmission or assumption of Hsia rites, since the chariot also flew the white flag of the Hsia.

The Shangs developed their feudal institutions in the valley of the Hwang Ho on the strength of their horse-drawn war chariots which were of similar type to those introduced about the same time into India, Iran and Mesopotamia, Syria, Egypt and Greece.

Horse skeletons, parts of chariots, bits, bridles and bronze fittings have been found at Anyang and other places. Shang horses were small with large heads and according to Watson[5] they could have been related to *Equus przevalskii*. They were driven in pairs.

As already mentioned in Chapter 1, one of the chief uses for keeping herds of horses was for their meat, although as Erkes says, apart from the fact that the Shang and also the Chou ate horse-meat, there is not a great deal of information on this subject. However China's western neighbours did indeed breed horses for their meat alone. In present-day Hsi-ch'eng-hsien (Chunghsien?), Sech'nan, 300 beef horses were offered to the Chou king Mu when he travelled to the western districts, whilst the Ts'ao-nu offered him 900, the Chi presented him with beef horses and others 700, the Hsi-mo 300 and the Ta Jung-hu 46.

Furthermore, on the banks of the River T'o, a personage by the name of Mu'Wang came across beef horses whilst milk from mares and cows was used by the Chü-sou to wash his feet. Duke Mu of Ch'in was not so fortunate, since the people of Yui on the southern slopes of the Ch'i Mountains which was the former home of the Chou people (before they invaded China), seized one of his carriage horses and ate it. The family name of Pang apparently signifies 'horse-slaughterer'. In later times it would seem that eating horses was no longer practised, although meat dogs are mentioned together with hunting and guard dogs.

There seems to be no doubt at all but that a great number of horses were bred in the provinces of China since only native horses were used in the first millennium B.C. for the many and varied invasions and wars, because, as the Chinese said, "they know the people of the country in which they are born". By the same token, a horse one has bred oneself is always easier to break in than a youngster bought in a strange market. These native horses were, of course, used only in harness, but frequent mention

[5] *China*, 1961.

is also made of their uses as "presents, bribes, tributes".

In 1125 B.C. Wu-yi, king of the Shang, presented ten horses to the Chou prince, and the western Jung of Tibet brought the Chou Hsiao-wang horses as tribute in 867 B.C. In 854 B.C. the Jung of Tai-yüan (Yangku) in north Shansi lost 1,000 horses after a defeat. Erkes further states that prisoners from both the east and south were entrusted with feeding the horses—so that even the eastern barbaric tribes were acquainted with horse management.

Thereafter, the breeding of horses on the vast grass steppes of the Chou kingdom was somewhat modified, quite probably because the land was needed for agriculture and for a growing population. Thus the north became the chief supplier and exporter of horses. "Horses are bred in northern Chi". Chi or Tai was called Tai-Ma-chün (*Ma* = horse, *chün* = district). The next-door state of Chin also bred horses for military purposes and the superiority of the Chin in battle was undoubtedly due to this policy. Around 283 B.C. the export of horses from Tai was reckoned amongst the three "most valued things".

"There are fast horses on the north coast [Gulf of Shantung]", writes Hsün-tse, and it appears that these were a breed of really well-bred horses.

Erkes sums up:

> Thus the most important centres of breeding were in central Shensi, the ancient home of the Chou, central Hopei, the kingdom of Ch'u which was the centre of the ancient southern culture and on the lower Yang-tse where, in the countries of Wu and Yüe the south-eastern culture had its centre and also in north-east Sechwan the seat of the Tibetan Ch'iang. So in all the old centres of civilization on Chinese soil well-bred horses were bred intensively.

The Chou-li gives three types of horses: war-horses, hunters and 'lesser'—or, as we would describe them, 'common' horses. Another source enumerates six strains or breeds: blood horses,[6] war-horses, parade horses, road horses, hunters and common horses, but probably the first applies to a nomadic culture who "valued their horses for war and hunting and those that were unsuitable for either were classified as of little value". Possibly the reference is to the Chou who were still partly nomadic when they assumed power over China. On the other hand, the six classified 'breeds' may have belonged to a settled society with an organized form of existence who had definite uses for suitable breeds of animals. Erkes, at any rate, is of the opinion that the first division is that of the Chou who had not entirely grown out of their nomadic habits when they conquered China, whilst the second could well apply to the Hsia who constantly used the numerical number six. The Hsia Dynasty commenced around 1900 B.C. and covered an area of 300 miles by nearly

[6] "first-class horses who resembled their dams". Chou Mu-Wang's harness horse Lü-erh (Greenears) was bred out of a black mare who came from T'ang in the north.

200, but soon spread over a larger area, until it was superseded by the Shang Dynasty *c.* 1639 B.C.

The ass was never a native of China—or, for that matter, any other part of Asia, being an importation at a fairly early date from Africa. But *Equus hemionus* was native. Its bones have been found dating to Palaeolithic times and it was hunted during the Shang Dynasty. It was described as "like a horse but with long ears", and is reproduced on Shang oracle bone and tortoiseshell. It would appear that the mule and ass became known through the Huns about 300 B.C., although the white mules which belonged to one Chao Ch'ien-tse of Chin are mentioned *c.* 500 B.C. Perhaps these were the result of horse × onager cross!

The rich and varied terminology which Chinese linguists have translated indicate that horse-breeding played an exceptionally large role. Most of these terms go back to the Chou period and if only we knew the locality of a particular dialect or expression we could establish the district of a particular breed or type of horse.

The following particulars are taken from Erkes, who in turn took them from the *Ku-chou-p'ien*.

chu, a horse with a near white hind fetlock (foot).

po, a horse of different colours; odd-coloured. Yin oracle bones.

wen, a horse with a red mane, white body and yellow-gold eyes. The *Shuo-wen* says that such horses were brought in by the Ch'üan-jung during the time of Chou Ch'eng-wang. On a tortoiseshell.

yin, a grey-white horse.

lo, a white horse with black mane.

liu, a red horse with black mane.

hsüan, a grey-black horse.

hsing, a red-yellow horse. This expression then generally used for sacrificial animals. (It seems possible that horses of this colour might have been specially bred.)

chui, a horse with a black-white skin (fur).

ch'i, odd-coloured; piebald or skewbald.

huang, a yellow-white horse (dun?).

hsia, a horse with a red-white fur (roan). Did the Hsia themselves specialize in horses of this colour? It is supposed to be the colour of very early primitive horses.

jou, a grey-black horse with varicoloured mane.

kua, a yellow (dun?) horse with black muzzle.

k'uei, a dull black horse.

liu, a red horse with black mane and tail (bay?).

yüan, a red (bay?) horse with black mane and white belly.

mang, a black horse with white face and forehead. Yin oracle bones.

p'iao, a yellow (chestnut? palomino?) with white spots or thick white tail.

yü, a black horse that is white between the legs.
ts'ung, a black horse with white spots.
hsien, a horse with one white (silver) eye.
yü, (fish) a horse with two white eyes (because it looks like a fish).
t'ie, a horse which is the colour of iron, red-black. (As Erkes remarks, this is interesting as it shows that the Chou knew of iron.)
yen, a horse with white hindquarters.
hsiang, a horse with off-hind fetlock (foot) white.
li, black. Yin oracle bones.

> With equal care both the height and age of the horses are given. The *Chou-li* describes an over 8 foot horse [15·3 h.h., 160 cm] as *lung*, one over 7 foot [13·3 h.h., 140 cm] *lai*, and one over 6 feet[7] [under 12 h.h., 120 cm] *ma*. And *ma* means simply horse. But in *Shuo-wen* this is also *chiao* for a horse over 6 feet.

Eberhard in *Lokalkulturen* says that during the Han period (200 B.C.–A.D. 200) there were horses that were only 81 cm (8 h.h.) high, no bigger than the smallest Shetland pony. But dwarf ponies were apparently bred and there are several references to small teams, dwarfs with carriages, horses and harness all of the appropriate size. Possibly they belonged to a travelling circus.

It was at Cheng Chou that archaeologists discovered amongst human burials the remains of one horse and five chariots from the Shang ii period, *c.* 1300 B.C. All were separate, the horse being about 40 metres distant from the chariots.

The royal cemetery (1300–1150 B.C.) at Hsi Pei Kang lies beyond the Huan River 3 kilometres north-west of Hsiao T'un. "One grave was filled with the dismantled parts of a chariot. Horses with head rope and bronze ornaments were buried apart, in one instance four together; in eight other pits in pairs."[8] At Ta Ssŭ K'ung, an intact chariot burial dating back to the eleventh or twelfth centuries B.C. was found. The skulls of the horses indicate the Asiatic wild horse which was apparently yoked to the chariot pole.

A burial of the Shang Dynasty (Period III) uncovered "the chariot's complement of four horses and three armed men". All these were sacrificial burials. In some graves the forerunner of the 'compound bow' was found and this was the standard type in eastern Asia as long as the bow was employed. "Its double curve," says Watson,

> affords a powerful thrust over a short pull, and in this respect it was ideal for shooting from the confined space of a chariot box. As far as we can tell the horse was not ridden in the Shang period though this type of bow was equally suitable for shooting from the saddle, and became in later times the principal weapon of the steppe nomads.

[7] It seems that a foot measures 8 inches and not the usual 12 inches.
[8] William Watson, *China*, Thames & Hudson, 1961.

Such a bow, however, could not be employed effectively from the saddle unless the rider had stirrups, and until the invention of stirrups, warfare was carried out by chariot. The invention of the stirrup loop strap was as much an advance as had been the earlier invention of the wheel.

At Anyang, Honaen Province, in the Ta Ssŭ K'ung burial already mentioned, the horses were yoked and thus pulled with their heads in an unnaturally high position. The next step in the Near East, followed by China, was bands round the horses' necks. Thus the harder the horses pulled, the more they tended to choke themselves and to carry their heads unsuitably high in order to run. "From the first century B.C. the Chinese possessed a form of harness which placed the strain on the chest and shoulders and obtained a much more effective draught. This was an invention which anticipated a similar European device, the hard collar, by almost a thousand years."[9]

So the order of draught, by which means the horses could be bred successively stronger and better, i.e. with more blood and speed, were: the yoke (which could only be worn by a plug!), choker bands, breast harness, collar.

Watson goes on the say that the Shansi and Shensi provinces of the Chou Confederacy were more suitable for horse raising that the Central Plain,

> and in historical times has supplied horses to the rest of China. It is also probable, although there is no archaeological proof of this as yet, that the Chou were acquainted with chariotry even *before* they moved east to conquer the Shang kingdom in 1027 B.C. . . .
>
> Chou chieftains were allied to the Shang from about 1500 B.C. to 1027 B.C. The resemblance of the Shang chariots to chariots made in the Near East towards the end of the second millennium B.C. is too great to be dismissed as a coincidence . . . If ideas of chariot design reached China from the West they probably traversed the Chou territory on their journey eastwards. It is conceivable that the Shang chariot was a borrowing from their western neighbours, on whom they may have relied besides for the supply of horses.

During the Chou Dynasty which covered the period between 1027 and 256 B.C., quadrigas of four horses were generally used. In a Chou tomb seventy-two horses were discovered together with twelve chariots.[10] The importance of a state was measured by the number of war chariots it was able to muster. During the Chou Dynasty, and this is not necessarily a contradiction to Erkes' statements but possibly ancillary to them, many foreign horses appear to have been imported for use in war and to improve local stock, Mongolia and eastern Tibet being the principal sources. According to Eberhard these regions possessed wild horses,

[9] Watson, op. cit.

[10] Creel, eighth century B.C.

superior domestic horses, particularly capable trainers and a tradition of horse mythology. Hančar believed that these were also the main sources during the preceding periods, but Watson says that in historical times the Shansi and Shensi provinces of the Chous supplied horses to the rest of China.

Both the ages of horses, by their teeth, and methods of training are known. Horses were kept in herds on large areas of grassland, as they are today in the Russian *tabun*. They were guarded by a herdsman or guardian whose job it was not only to care for the horses but in spring to burn the grass to improve the quality of the grass, and to see that the mares were covered. There were twelve royal studs which were called *hsien*. In 1900–1300 B.C. a whole district was called Hsien. These royal studs were under the direction of a stud-master who had to control the fodder and to see that the stallions were fit and well. Colts were gelded at two years, as in most modern European breeds, and horses were broken in at three years old. Two-year-old horses' ears were pegged with bamboo clips and this was supposed to have a subduing effect. Horses were taught the six gaits by the breeders themselves. These were (after Wan-Yung-tien quoted by Biot): walk, halt, lead, step back, trot and canter. The animals were fed according to their work; they were not overtired, and they were properly housed.

The earliest veterinary surgeons, or people who cared for sick horses, were probably women *wu-ma*, but there were also horse-valuers, *Ma-shi*, who fixed the price paid by the Chou state for horses.[11]

"Thus," says Erkes,

> this is an indication that besides the state studs there was an intensive private breeding of horses. The valuer was a specialist, since he told the age [by the teeth], coped with bad horses and made compensation for the loss of a horse through early death, assessed the weight a horse could carry and acted as judge in all differences of opinion concerning horses.

These valuers or specialists came, according to Eberhard, from Chin and Chao. Thus breeding and knowledge of horses were most developed in the ancient Hsia country which again proves that the oldest centre of horse-breeding was still in the oldest district of Chinese culture.

The main objective in breeding was naturally for military purposes and the title of the Minister for War himself was *se-ma*, Master of the Horse, and surprisingly, perhaps, it was customary from very early times when camp was made that it should be surrounded by waggons or chariots—the waggon fortress as it was called.

Horses were bred then for meat and for a variety of uses in harness but they were not bred as riding horses. They were also not ridden in India or

[11] Presumably the Chou Dynasty, eleventh to ninth centuries B.C.

Egypt and the Hittites who had come to power in Asia Minor at the beginning of the Hsia Dynasty also used only harness horses for the army (see Chapter 4). It was about 307 B.C. when, according to Eberhard, the "King of Chao introduced riding together with the use of the bow by mounted horsemen and the dress necessary to go with it, from his northern barbarian neighbours." The Yen family name of Ch'i: rider, appears around 300 B.C. Erkes thinks that riding was unknown even to the northern barbarian tribes only 200 years earlier. This places riding in the Far East or Far Central North East at around 500 B.C. Whether the date is really quite as late as this it is difficult to say as it would depend at what period nomad steppe tribes copied the habits of their forest tundra neighbours who had *certainly* tamed the reindeer to ride at a considerably earlier date. Then again, the steppe pony has to be big enough to ride even accepting that the men were smaller at that time. However, when armies were mounted—and this could not have happened all at once—instead of chariot-driven, they rapidly became more versatile and consequently more powerful.

Whilst the various provinces had successfully bred local types of horses, some probably proceeding from beef horses to pack horses to harness horses, when it came to good riding horses which had of necessity to be of heavier conformation, the attention of Chinese emperors was drawn to the West. The changeover from war chariots to mounted warriors was essential in order to combat the mobile mounted nomad barbarians from the north and west.

In about 166 B.C. the Hsung-nu (Huns) of what is now Mongolia invaded northern China with, it is said, 140,000 horsemen mounted on a heavy type of horse acquired in the West. Between 138 and 126 B.C. Chang Ch'ien journeyed from the Chinese capital to Yüeh-chih which lay to the west of the Kunlun and Pamirs, and he was the first to inform the Western Han Emperor Wu-ti (140–87 B.C.) of the "blood-sweating[12] heavenly" horses of Tayüan on the Upper Syr Darya (Jaxartes) River in Turkestan. Tayüan lay on the eastern outposts of the Hellenistic world. Since Tayüan lay on the other side of the Tien Shan Mountains (or the Celestial Mountains to translate literally), it does seem probable that the Chinese thought of them as "celestial" rather than "heavenly" which is possibly a Westernized expression and has been used by many writers to mean a super breed. I do not think this was so. The Celestial horses from the other side of the Celestial Mountains were good and better than most Chinese horses, although we must not forget "the fast horses on the north coast" of China.

[12] Blood-sweating of horses is due to the parasite *Parafilaria multipapillosa* which is common in Central Asia. Cases have been known in Britain. Tiny 'blisters' are caused in the summer months by the parasite under the skin and when broken through pressure (of saddle) or activity, the blood mingles with the sweat.

In many parts of Europe herds of wild horses existed into the eighteenth century whilst a few of the large forests, such as Rheinhartswald in Germany and the Fens in Britain, provided sanctuary to a later date. There were Tarpan on the steppes of southern Russia, it is said, even up to 1925—which if true confounds the widely held belief that they became extinct much earlier. There is the, to all intents and purposes, wild Sorraia horse of Spain and Portugal and the Garranos of northern Portugal. Both breeds have been extensively investigated by the late Señor Ruy d'Andrade who published material on the subject. Otherwise these breeds are unknown to or have been ignored by European writers. There are remnants of herds of the Asiatic wild horse still, we hope, to be found on the southern slopes of the Altai Mountains, and there are said to be wild horses in far northern Siberia in the vicinity of the River Kolyma. There may still exist in remote areas small herds of wild horses that are unknown to the Western world.

In order to bring some relevance into this history, one must choose from the historical material available an approximate date when selective breeding was practised and strains of horses differing from each other were evolved. In Asia such selection soon had as its main objective breeding for military purposes, commencing, of course, with nomadic expansion. (The Asiatics seem to have developed very early on a certain territorial aggressiveness.) In Europe horses were wanted primarily for mobility and only secondly for skirmishes. Hunting and fighting were man's chief occupations and the horse soon became his most important accessory.

It is surprising how rapidly the numbers of horses increased from the Yellow Sea land westwards to the North Sea and the Atlantic Ocean. Almost every tribe and every people between latitudes 60°N. and 25°S. from about 2000 B.C. on, right up to the present day, were engaged in furthering the evolution and domestication of the species they most admired and needed. The Arabs consider the horse follows man in the order of Creation.

Climate and food i.e., the mineral, protein, and vitamin content and quality and quantity available, play a most important part in the development of a species, not only in the outward physical development but also in the make-up of genes including the factors of inheritance and the fixture of a type and/or mutations.

Basically the following conditions 'breed' certain types of horses:

1. Desert/grass steppe produces a light wiry animal with stamina.
2. Mountain/lush valley produces a short-backed, short-legged, sure-footed hardy animal.
3. Chalk/lucerne grass[13] areas produce a larger, possibly heavier, well-bred animal.

[13] It should be borne in mind that 'grass' covers many and varied sorts of herbage.

4. Islands produce smallish animals after the pattern of the first introductions.

From the natural living conditions in the wild which were normal for the species for hundreds of thousands of years, when certain types evolved, the horse responded to 'domestic' conditions which covered only a few thousand years during which evolution was speeded in varying degrees and directions. Some types retained their original characteristics of a pattern that is so fixed that it influenced others whereas other strains of less determined conformation characteristics have become merged or lost altogether. Strains have vanished and new breeds have appeared.

In the wave of wars, invasions and the dissolution of peoples that have taken place during the past 4,000 years it is not easy to follow in most instances an exact line for more than a few hundred years. Although peoples disappeared or merged with their conquerors, the strain of horses general to the country in particular may continue in the same district, may improve or deteriorate or may suddenly be found in quite another part of the world, as, for example, the modern Arabian horse in the United States of America.

Some hippologists think that the heavy horse evolved in Europe, others that Asia and particularly Persia can be ascribed as the original source.

The size or height of the horses used in early historical times can, to some extent, if the bones are not available for measurement, be judged by the width of the mouthpieces of those bits that have been found in Europe and elsewhere—from the antler bit from the Heathery Burn caves in County Durham, the curved and perforated bit cheek-pieces of horn from Terramaricoli in the Po valley *c.* 1500 B.C. Horn bits were specifically a European invention and the date 1650 B.C. is the earliest proved occurrence for the horn cheek-piece found at Tószeg in Hungary. Later both bronze and iron bits of the Hallstatt period indicate in many instances that they were made for and worn by ponies rather than horses. But in Europe there is very little information about breeding, how ponies were bred or where they were bred, with the exception of the discoveries near the Dümmersee in north Germany which appear to indicate that horses were actually bred there *c.* 2500 B.C.

Professor C. F. C. Hawkes in *The Pre-historic Foundations of Europe* says that the corded-ware warriors of North Central Europe and the Swedish boat-axe people (*c.* 2000 B.C.) are proved (*sic*) horse-breeders, and that the breaking of horses of European breed grew over Europe until in the Late Bronze Age they became a potent instrument of domina-

---

There are special grasses suitable for horses and others for cattle. Certain so-called weeds are also necessary in the diet of both. Thus some areas or localities of the earth provide more or less nourishing fodder.

tion. I asked Professor Hawkes for this proof but I have received from him no further reply.

Somewhat earlier than this date, the Danubian colonization near Lindenthal in Germany built oblong barns on stilts in which to store their crops, and enclosed in ring ditches were open spaces for flocks and herds. During the Cucuteni Tripolye culture at Izvoare, Bistritza River, Moldavia, horses' bones were found recently amongst those of sheep, pigs and cattle—an indication that the horse was domesticated.

During the last glacial period, 10,000–6000 B.C., there was a great decline in wild horse populations and large and small horses lived alongside one another for only a short time. Both Mesolithic and Neolithic horses were similar to *Equus przevalskii przevalskii* Poljakoff e.g. 13·3 h.h. (140 cm) so that possibly the Przevalsky horse was the first to be domesticated. Lundholm insists:

> The earlier assumption that the Przevalsky horse existed in Europe during the Ice Age is erroneous.[14] The western group can probably be divided up into two forms of different size. I have called the larger the 'Germanicus' group and the smaller one the 'Microhippus' group. The bio-geographic conditions during the Ice Age explain the division of the wild horses into two groups. The region between the Nordic ice of the Alps lay above the dwarf-scrub heath limit and was in winter covered with great masses of snow. This region, which in summer offered pasturage to grass-eating animals, was in winter entirely evacuated by these animals. The horses in the west probably migrated[15] down to the woods to the south-west of the Alps, and found there a meagre winter pasturage. Those in the east were forced by the mountain regions south-east of the Alps to migrate due eastwards until the depth of snow was so slight that they could get at the dried grass underneath.
>
> When propagation took place in the late spring or early summer the groups were still isolated . . . The west group lived on forest pasturage, while the eastern group grazed upon the dry grass of the tundra steppe. This has affected above all the development of the dental character, the structure of the mandibles and the bones of the extremities. Wild horses existed in various places in Europe into historic times, including Gotland [small in stature] as in other island forms.

During the Neolithic period the horse occurred sparsely over practically the whole of Europe. From the Stone Age to the Bronze Age there was a marked increase in horse finds indicating domestication which was illustrated by pictures of horses, bridles and chariots. From the dis-

[14] Professor Herre of Kiel University would seem to be of another opinion. He says (1967) that fossil and sub-fossil documentation shows that once *Equus przevalskii* was widely distributed in Eurasia north of the Caucasus. Vast herds of these animals were to be found from West Europe to East Asia.

[15] Lundholm says "probably". I find it difficult to believe this to have been a voluntary form of migration. If pasturage is available in winter, it will certainly be available in summer, and horses, as we have seen, like to remain within a known territory.

coveries we may assume that the horse was domesticated more or less simultaneously over great parts of Europe.

Munro (1902) writes about Neolithic horse remains in France: "The osseous remains indicated two varieties of the horse, one large and the other small but they showed no evidence of domestication." There is evidence from Bohemia of metacarphalia of the Neolithic period showing that these horses were small with fine bones.

What is domestication? Lundholm gives the best answer (translated from the German):

> Domestication is a symbiosis between people and animals which has lasted through many generations. This definition therefore excludes any accidental taming of odd animals. To be exact, it need not necessarily mean that the animals are tame, only that throughout a long period they have had a dependent relationship with man. The definition also means that man must have had a certain use from the animals.

When wild horses submit to even a simple form of domestication certain changes begin to take place but it is by no means always possible to tell if fossil remains are those of wild horses or of domesticated horses. One of the most noticeable changes takes place in the teeth and in the brain cavity. Furthermore, conformation changes also take place—some horses could acquire broad foreheads or long faces and others short faces and convex or concave profiles. According to such conditions as climate, selection—or lack of selection—in-breeding, and fodder, horses can diminish in size and small pony types may develop. In the same way large, heavy horses can also evolve and these changes are the result of domestication in a slight or exaggerated degree. As we have seen in some of the recent modern breeds, even changes of head carriage and gait can be developed. Domestication requires a different use of brain and the organs of the senses, which in some cases is less and in others of a much higher capacity. In situations to which a wild horse is a stranger, a domesticated horse must react and then respond accordingly. And then there is no doubt but that a systematic form of selection has been exercised to breed horses with good "intellects".

"Dwarf forms of horses," says Lundholm, "often appear in conjunction with primitive domestication." These are met with during the Bronze Age and very often such dwarf forms appear in very isolated areas not connected with the geographical situation, such as islands. To take instances distant from each other, we have the dwarf Skyros pony of Greece, the tiny pony types of Timor, Bali,[16] and Indonesia, and the well-known Shetland pony from the Shetland Islands. Diminution may happen very quickly, possibly inside a few generations, whilst it takes much longer to increase size and weight. In both cases, undesirable char-

[16] See Daphne Machin Goodall, *Horses of the World.*

acteristics may develop caused perhaps by the appearance of a peculiar gene—mutation—and this may soon influence an entire population or breed. It can be 'bred in', whereas another which might be of great benefit—an improver—can easily disappear or be 'bred out'. The one gene has become fixed, the other not. If the variation frequency to begin with is 50 per cent, the probability of fixation or disappearance is equal. If the frequency is less than 50 per cent then the probability of disappearance is greater than for a fixation, and inversely.[17]

Once a gene has become fixed then only mutation can change it.[18] In the case of wild horses or wild animals of any species, undesirable characteristics may be bred out in so far as these characteristics make the animal incompetent to fend for itself.

During domestication there appear also a large number of different mutations. These are presumably caused by a heightened mutation frequency in connection with domestication. Probably disturbances of the hormone system are the fundamental cause. Lundholm considers that these disturbances lie in the changes of the sexual rhythm occurring in transition to the domestic state; this may be caused by change of diet and particularly the physical exertion which domestication entails.

For a detailed analysis of mutation as well as the *science* of the genetics of breeding, the reader and serious breeder is referred to *Genetics of the Horse* by Dr William E. Jones and Professor R. Bogart (1971). Briefly, "The genes in a multiple allelic series are transmitted in exactly the same manner as other alleles. It has been postulated that there are multiple alleles in many performance traits . . . the best examples of multiple alleles in horses are found in the blood types and transferring types."

Further, "Certain basic principles should be understood about mutations. Genes may mutate, for example, from A to *a* and also from *a* to A . . . In a species such as horses where the rate of production is extremely low one should not expect to see many new mutations." The authors then sum up the causes for mutations: ionizing radiations, the alpha, beta and gamma radiation of radioactive substances and X-rays, ultra-violet rays, high temperatures, chemicals and drugs. Of the Tarpan (*Equus przevalskii gmelini* Antonius), Lundholm has this to say: "The tooth-form diagrams show that the Tarpan must have been the progenitor of the eastern European domestic horses," and further,

> Lower jaw characteristic would seem to show typical 'domestic horse lines'. If one considers the Tarpan to be a genuine wild horse, then it would have to mean that the characteristics of the lower jaw were once common to wild horses.

[17] S. Wright, *Evolution in Mendelian Populations*, 1931.

[18] E. Fischer, *Die Rassenmerkmäle des Menschen als Domesticationserscheinung*, 1941.

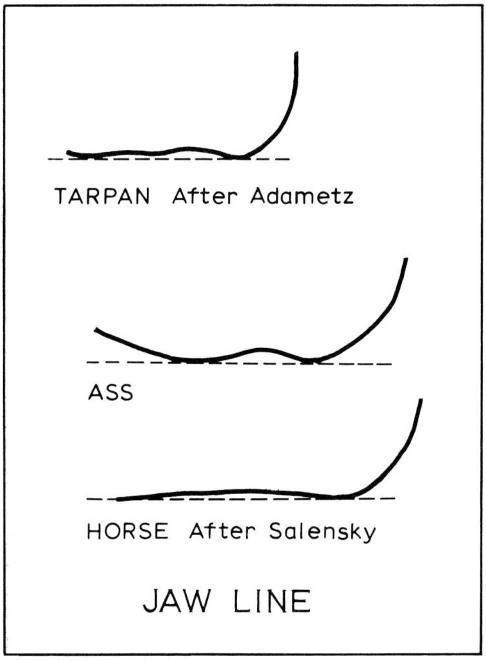

28. Diagrams: Comparative jaw lines of Tarpan, Ass and Horse; and head profiles of Barb, Arab etc.

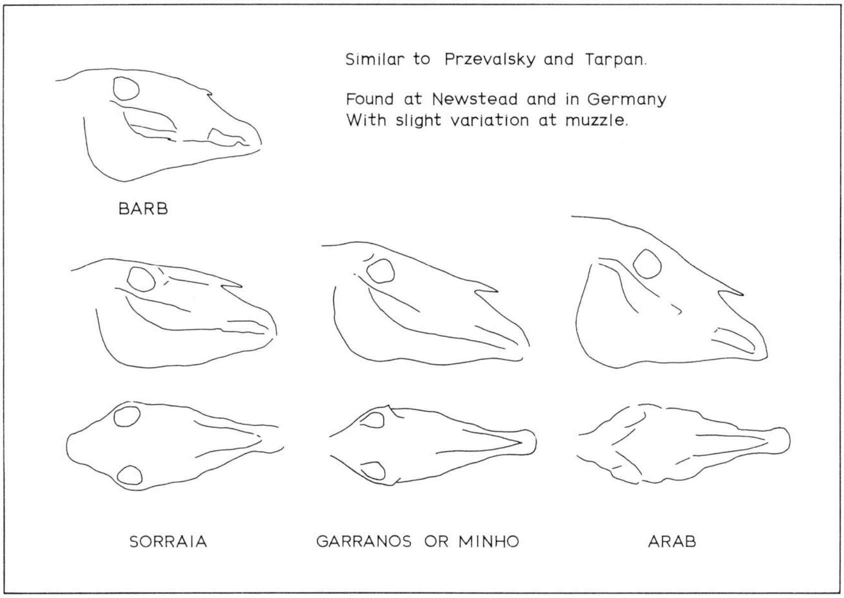

In adult horses, obviously the length of the teeth have a bearing on the shape of the lower jaw: a straight profile means that the teeth are shorter and it is not difficult to appreciate that modern horses have shorter teeth because the diet of domesticated animals is not as demanding on the teeth. Equally, it is strange, that the wild Tarpan does not need to have such long teeth as other wild horses.

The general opinion is that most domestic horses are descended from the Tarpan. [Author's italics.]

First of all we must be clear what the Tarpan is and what we actually know about this horse. For a considerable time arguments have raged as to whether the South Russian Tarpan was really a wild horse or only a feral horse.

Gmelin (1770) first described it as a wild horse. Pallas (1831) shared the same view, as also did Schatolow (1884) and von Falz-Fein (1919) who came in direct contact with the horse. More recently above all Antonius in 1912 and 1922 especially drew attention to the true wild characteristics of the animal. But this opinion is contradicted by a number of Russian explorers who must have been acquainted with the habitat of the horse. Anutschin, quoted by Nehring (1884), that it is only a feral horse. Also Bogdanoff (1926) and some of the English experts like Ewart and Ridgeway. Kwaschin-Ssamarin (1931) who occupied himself extensively with East European primitive horses, contends that "there is no doubt but that the Tarpan shows signs of domestication. So the question arises, if there ever were wild horses who had the same characteristics as the Tarpan".

Lundholm, however, sticks to the opinion that the Tarpan was a genuine wild horse and I need hardly repeat that I am still of this opinion as far as those regenerated animals are concerned which today are found in Polish reserves but not, I must stress, as far as the so-called Tarpan which Heck has bred in Munich Zoo. These do not show the same *Tarpan* characteristics as those in Poland.

"The conformity of the form of the diagram teeth of the Polish konik, the Tarpan, the Arab horse and the wild horses of Kasan all agree." (Antonius) That the Arab horse belongs to the East European group has already been emphasized by Adametz (1926) and Hilzheimer (1925). Pavlova (1890) produces evidence that there were undoubtedly wild horses with the skull characteristics of the Tarpan, in his discoveries in the Kasan Province. These finds were found together with rhinoceros, ox and deer, so that the genuineness cannot be doubted.

The evidence before us is, therefore, that the Bronze Age inhabitants of Europe were able to domesticate a small type of horse of about 12·2 h.h. and a larger type of 14·2 h.h. Some of these one would conclude to have been of the Przevalsky strain whilst the majority will have had a large element of Tarpan blood and this element we may be able to follow down the centuries by means of illustrations.

At the time that some sort of organized form of domestication of horses in Europe really began to get under way it seems probable that there were two types—or, perhaps, strains and probably local varieties of these types. As the basic originators, we have the European form of the Asiatic wild horse (Przevalsky) and the Tarpan. It is my suggestion that the Tarpan, *Equus przevalskii gmelini* Antonius, provided the basis for *Equus europaeus*, the so-called plateau horse, which fathered in a less mixed form the later konik strains of Eastern and Central Europe, bearing also in mind that island strains would diminish in size; and the Sorraia of Spain. When Tarpan crossed with the Przevalsky horse, the Exmoor pony evolved, otherwise it is difficult to explain the extraordinary fixed characteristics of mealy eye, nose and belly.

Both the Exmoor and the Sorraia became isolated as types or strains in later prehistoric times. Many caves in southern France and northern Spain and Portugal have produced the most splendid murals, carvings and engravings. Some of these point to an earlier contact or domestication of the horse than we have so far been able to discover in North and Central European societies, and, by means of the then-existing land-bridge, there could easily have been an equine connection with North Africa. Possibly this took the form of dual distribution from north to south during glaciation and from south to north as the ice receded and the climate became less severe when tundra conditions changed to steppe grazing lands. The isolation of the Exmoor took place most likely by 6000 B.C. when the land-bridge joining Britain to the Continent was flooded over by melting glacial ice from the North Sea and the English Channel became a water-way.

The Sorraia was isolated eventually by the nature of its steppe–pampas habitat in central Spain. This is not the place to speculate that the exported native horses of Spain, to South America, later known as the Criollo, acclimatized so well because its habitat was a 'home from home'. We shall come to this in Chapter 7.

I also suggest the cross Przevalsky × Tarpan laid the foundations for *Equus robustus* typified by the present existing cob types: Ardennes, Fjord, Hafflinger, Mierzyn, Breton, Welsh, etc.—who in turn provided the foundation for the centuries-later heavier horse breeds.

In Asia, there was *Equus orientalis*, whose modern descendant may be found in the small local pony of the Elburz Mountains, commonly called the Caspian pony.

There is very little evidence for a widespread domestication of the horse in Northern Europe during Neolithic times. Indeed, the discoveries of the hunter/cattle-breeding settlement around the Dümmersee show only three types: a small domesticated horse with inferior bone, a rather larger wild form, and the coarse-legged *Equus germanicus*.

The second archaeological evidence for Neolithic domestication and

29, 30. Horse designs in metal, Egyptian New Kingdom, 1600 B.C.

31. *(bottom)* Trundholm chariot, Denmark, 1400–1200 B.C.

breeding of horses in Central Europe came from Jordansmühl, Złota (Poland) and Potyry. The third comes from Bodzáspart (Badener culture), where part of a bone bit together with a number of horses' bones were found.

These later Stone Age/Bronze Age peoples of Europe will, as I have already shown, have kept their 'domesticated' horses in herds in a semi-wild state, where they bred on the open grasslands and were enclosed only when needed for religious or ceremonial purposes or in times of unrest. It is not proposed to produce further 'evidence', suffice to say that from about 2400 B.C. onwards, bones of horses, bits and models of carts or engravings have come to light. As folk learnt more about bitting and harness (except for the yoke, neither being necessary for draught oxen), the horse took over as a means of transport probably from the onager via the ox and reindeer. The earliest cart was the triangular sled to which the reindeer was harnessed and also ridden, and the horse may also have been ridden at first in the shafts of the sled!

The Donau area especially would seem to have been the embryonic centre of horse-breeding. Perhaps the idea stemmed from the Beaker folk, who seem to have been both inventive and energetic and explorers as well. At any rate, they acted as a bridge between the Neolithic and Bronze Ages and spread the knowledge of metal-working through Europe to Britain and even eastwards via southern Russia to China. They were succeeded by the long-barrow cattle-breeding folk and by 1250 B.C. the momentum of resettlement of urn-burial peoples from central Germany and Hungary had spread in all directions, particularly to the north, west and south where it reached the Hatti kingdom and Egypt. During the second millennium peoples were on the move either through pressure or of their own volition. To move anywhere at a distance, one must have faster transport than oxen provide. Thus the knowledge of horses, the uses to which they could be put and methods of training were spread.

We should remember that cattle were not 'improved' in Britain until the advent of Robert Bakewell (1726–95) just on two hundred years ago. He was the pioneer breeder who improved Longhorn cattle and also bred working horses and generally started the improvement of livestock in Britain. Even then there were no breeds in the modern sense and a considerable variability occurred in local types, "ancestry was mixed and transmission apparently very irregular".[19] So it becomes obvious that our European forebears were not concerned with genetics as such, although the pattern in Asia was perhaps more selective since chariot horses were required in large numbers with increasing necessity.

In Europe it seems likely that bad, lame or unreproductive horses would be culled for the pot, and as likely, too, that a good mare would be

[19] Watson and More, *The Science and Practice of British Farming*, 1942.

kept and bred from, but there is not a shred of evidence that any form of controlled selective breeding was known or practised.

For many hundreds of years the same sort of animals appear in the art of the Greeks or the Celtic peoples. This was the smaller horse used in the chariot and, as time passed, the rather larger riding animal. A kind of universal strain; hardy, easily manageable whether in harness or ridden.

With the great number of afforested areas in Europe and in northern Russia in addition to the prairies or rough grasslands, it seems likely that, following the use of the horse for pack transport, riding was first practised here and spread eastwards. It is always easier to ride through wooded country than to try to drive through it. Conversely, I think the use of the horse in harness as already mentioned, especially in the Donau area of social progress, began as a sequence to draught oxen which were used by almost all nomadic tribes to transport their goods on carts. In those far-off days, the tracks in mountainous districts could only be used by individual animals so goods had to be *carried* rather than *pulled.* In order to work in hilly or mountainous country a horse needs to be strong, short-legged and robust both in health and disposition—*Equus robustus*—and it was, I am convinced, in such districts that the cob evolved even if today the type or breed is found in other areas. But one thing is certain, no one associates the cob with desert, pampas, tundra or blue grass country!

We have to keep before us the recognition of climatic changes and therefore vegetation growth in respect of the possibility or otherwise of breeding horses and that such events bear on their development or diminution. One of the most important things that happened in post-glacial Eurasia was the increase of climatic characteristics to the optimum of drought and warmth in the third and second millennia and the worsening conditions towards the first millennium. Professor Hančar goes on to say that the Tripolje civilization at the time of its early horse-breeding activities is the most thoroughly investigated amongst the incredibly varied mosaic of societies of the East European steppe or forest zones. Already in 3000 B.C. the earliest Tripolje folk had connections with the Beaker folk who were already well established in Central Europe. The former occupied the Ponti steppes (fourth millennium) around the Dnjepr and other rivers Desna, Oster, Bug and Supoj (Ukraine), which country was in later centuries to provide a home for a number of displaced horse-loving peoples. The Tripolje people, who were primarily interested in fishing and hunting, lived in houses in village communities. About twenty to thirty people lived in each house, which was equipped with ovens, burnt clay floors, painted walls, living-rooms and pantry. One thousand settlements have been discovered.

The Tripolje civilization divides into three chronological stages: Period A, 3000–2700 B.C.; Period B, the heyday of the Tripolje civilization, 2700–2100 B.C.; Period C, 2100–1700 B.C. They grew corn, which was harvested with flint or bone sickles; millet, wheat, barley and rye, and the textile 'industry' was still in production in Roman times 2,000 years later! They bred large and small horned cattle (*primogenus* and *taurus*), sheep, pigs and goats, and they had dogs and the more rare horse. They were also one of the early historic societies to have a matriarchal form of society.[20]

Their woods, rivers and parklands were inhabited by no less than eighteen species of wild animals: deer, stag, elk, wild boar, badger, gula gulo, pine and stone marten, otter, polecat, wolf, fox, wild cat, lynx, hare, squirrel, beaver and hedgehog. Bones of all these animals have been found at Luka-Vrubleveckaja, and in addition at other places bones of the aurochs, saiga-antelope and onager show the tremendous variety of wild life as well as domestic animals with which the Tripolje folk were daily surrounded.

The country and the environment must have been a veritable paradise. Indeed, ancient peoples may have recognized this and thus the legend arose that mankind was born in Paradise! However this may be, the fact is that as man has travelled down the centuries, he is now both morally, spiritually and physically a long way from the natural paradise of his origin. In time, the distance from those days is short and it occurs to me that we may not realize that the heyday of civilized man could already be over.

Apart from the Tripolje civilization when horse breeding began pre-3000 B.C., the Jordansmühl Beaker people also bred horses *c.* 2500, whilst earlier in northern Europe, the period of dolmen building, probably horses were being bred.

But the second civilization where horse breeding has been established lies in the Ural and Lower Ob district of the south Transural Aeneolithic Afanasjevo society in the Siberian forest-steppe. All had knowledge of the principle of cattle breeding whilst the Ural–Altai district had close connections with the Turan wild ass or onager breeding—Anau; and even further back in time to the reindeer-breeding peoples who, like the Jordansmühl and Tripolje folk, had contact with wild horses. It is still an open question as to whether the south-east forest-steppes of Europe should be acknowledged as the source and origin of the domestication of horses. It seems very probable that this knowledge spread from the Tripolje centre north-westwards and eastwards, particularly as similar ceramic forms have been discovered in these two areas.

This knowledge appears to have come to the British Isles around 2000 B.C. when horse breeding is considered probable—it certainly existed

[20] This is interesting as a line of thought. Many horse-loving peoples enjoyed a matriarchal society, and, omitting the legendary Amazons, such societies generally lived in peace and propagated an active civilization which lasted many centuries.

about 1800 B.C. in the early Bronze Age. This is also the approximate date given for the building of Stonehenge where, possibly during ritual festivals of the sun, chariot-racing is supposed to have taken place. During the fourth century B.C. Pytheas writes of the temple of Apollo which is erected on the island of the Hyperborei[21] and which possibly refers to Stonehenge.

It may have taken up to two thousand years for the horse to be looked upon as general-purpose transport in place of the ox. The ox is easily harnessed by means of the yoke which lies easily between the horns or on the neck, whilst the yoke, although used, was quite unsuitable for horses as a means of draught. The next stage was the choker and although used and frequently seen to be used it was not very satisfactory either—partly for the reason that the choker was not only attached to the centre pole to hold it in place but the same line or thong was used to pull the vehicle. No outside trace appears to have been thought of. In modern pair harness, it is the traces which take up a high percentage of draught.

The Sumerians *c.* 3000 B.C. had already developed the *idea* of riding with their invention of a two-wheeled vehicle with a seat which the rider/driver sat astride. This vehicle was drawn by two onagers since, as we have already seen, the horse was a rarity and as Hančar writes there is no word in either Sumerian or Semite for 'riding'.[22] And even if anyone had the desire to sit on the backs of such small horses as *Equus orientalis*, who would appear to have been about 10 h.h., such a practice was impossible. The horses were small with slight bones and weak backs.

The first example of a man on horseback is the bone engraving of the Susa Horseman (2800 B.C.). But the engraving is sufficiently indefinite as to make no distinction between the sub-species *Equus*: horse or onager. In Europe and Siberia horses were stronger and even bigger and people had the tradition of riding reindeer.

Progress in developing the draught power of the horse, however, came from Asia when, finally, breast harness was invented. Then man became master of the first effective instrument of war, invasion and destruction, and leaders rose to positions of power. With the faster method of transport men turned to arms, armies were horsed, kingdoms were conquered, states expanded, people enslaved and policies made. And the first warrior knights appeared as a class. The movement reached its zenith during the second half of the second and first millennia B.C.

The great and peaceful state of Sumer-Akkad was at the mercy of any local invading army, any mounted bandit tribe bent on destruction—what else were the Guti who destroyed it?

[21] '*Hyperborei*' means in Greek 'beyond the wind'. They were a legendary people of the far north who enjoyed perpetual youth, constant sunshine and unclouded happiness.

[22] Although later the meaning of 'driving' included also 'riding'.

Thus the equestrian states of the Near and Middle East, of the Hittites, Mitanni and Cassites and even Thracians[23] came into being. The power of these kingdoms was based upon the war-chariot. Later on Babylonia, Assyria and Egypt overran them with huge part-cavalry divisions; and later still the Medes, Persians and all the nomadic tribes that came under the Scythian umbrella spread far to the west. The Scythians (Greek for nomad) were succeeded by the Hun and Magyar invasions, although the Huns flowed over China before turning westwards.

On the European scene, it is not until we reach Thrace that we find any serious use of the horse—and therefore horse breeding—for war. Pre-Thrace itself had suffered, as early as 3500 B.C., from invasions from the Near East and there was a continual Aegean–Donau, Anatolian–Donau interchange, until in 1800 B.C. we find the early Thracians and early Greeks consolidated. And here again domestic horses were required in numbers and so far as one can tell the horses were similar in conformation and type, whether they came from the north—the Tarpan-type sun-chariot horse *c.* 1400 B.C., or the rather late seventh century B.C. illustration (page 127) from Italy, or indeed those of the determined band of horsemen portrayed on the silver bowl (page 128) of the first century B.C. All these representations seem to be of the *Equus europaeus* 11–12 h.h. strain, a generally useful animal which could be ridden and driven.

To sum up, the Tarpan, *Equus przevalski gmelini* Antonius, was widely, although perhaps not profusely, distributed in Spain, Central Europe and Asia including the southern Russian steppes. It was, and is, distinguished especially by a broad forehead with comparatively short face and domed appearance of the outer cranium wall. Adametz also says there is an inclination to lose the chestnuts on the hind legs and that the Tarpan has five vertebrae giving a compactness of the rump. The colour is always more or less mouse dun with dark dorsal stripe and sometimes striping of the legs. "Descendants of the Tarpan are the improved country (native) horses (even to atavism) of East Galicia, Russia and the east Balkan states and the moor-sand district of Bilgorai where Tarpan descendants are widely spread. The strain is further spread throughout Bosnia, Albania, Macedonia and Greece."

The Tarpan is represented often by diminution in many island ponies, for example, the Skyros, Lofoten, Gotland, etc.

It should, perhaps, be mentioned here that in the early part of the twentieth century a number of writers on hippology and scientists were firmly convinced that the Tarpan was a feral animal and that it was descended from a wild *oriental* ancestor, was domesticated and then be-

[23] It would appear that the pre-Thracian people or forebears of the Thracians were late Tripolje folk—which accounts for the unbroken knowledge of horses.

came feral. This opinion is still held by some people. In view of this present research, it seems more probable, and indeed the evidence has been produced, that the wild Tarpan *was* the ancestor of the wild oriental and European strains. Some became domesticated in the course of time and some remained in the truly wild state, whilst some may have been domesticated by early man and then returned to the wild state. This, at least, is the opinion I have reached after some years of thought and research amongst quantities of material. I therefore conclude that the Caspian pony or Elburz Mountain pony and the many strains of Turkmene and later Arabian horses are descendants of the Tarpan, although in some cases that descent may be no more recent than 10,000 years, or *even longer.*

We are dealing with hundreds of thousands of years of evolution and it is only within the past 5,000 years that we can begin to type horses into the beginnings of probable breeds or strains.

The Przevalsky horse (Asiatic wild horse), *Equus przevalskii przevalskii* Poliakoff, leans towards a slightly convex face with a certain lengthening of the muzzle when compared with the Tarpan. "There is," writes Adametz,

> a definite slight approximation to the occidental type of horse. It clearly heads a line of equine forms which connect with *Equus germanicus* Nehring (*Equus robustus*). It has five vertebrae, height is about 13–13·2 h.h. (132–137·2 Stm). It is of even compacter conformation than the Tarpan, the extremities are stronger with more bone. Colour varies according to locality from dun to red bay, dorsal stripe and zebra stripes occur. It is the only living wild horse which provides the basis for certain domestic forms.

Some experts maintain that the Przevalsky, too, is feral, although of ancient origin. The reason given for this opinion is that the locality in Dsungaria where they are presently found, which is today more or less a desert, which "from the ruins to be seen was once a widely populated area and thus could not have been inhabited by wild horses" [Adametz]. Sinkiang (Hsin-Chiang) is, and of course always has been, a land of great mineral riches. It is subjected to tremendous sandstorms and sand has covered what were once towns of importance.

Although contradicted by some writers, the Przevalsky horse was also spread as far as Spain and France and Adametz refers to the "wall of bones at Solutré" as evidence that it was hunted for its flesh. I would suggest that this is an error in logic. Early man would not have eaten his game *on the spot*, thus the great number of horses would not have accumulated. Solutré, I think, was a disaster caused by an unnatural element. But the Przevalsky horse's centre of domestication "lies in central Asia" (long ago he was driven out or became extinct in Europe) where since the beginning Asiatic peoples have domesticated it, and its descendants are widely spread as Mongolian, Manchurian and Chinese ponies (but not south Chinese) of different strains. And it is from this centre of domestication that we shall continue to follow his trail and that of his neighbour, the Tarpan and his descendants.

The culture of the nomad herdsmen, which still exists, has been a way of life far back into prehistoric times. We must not imagine that civilization began with the earliest written evidence, or even with the earliest pottery. It was a fact 20,000 years ago—Altamira, Lascaux, etc.; 30,000 years ago—Tübingen ivory in-foal mare; and 70,000 years ago—Shaffhausen 'commando baton'.

A people living within a geographical area are known *to us* under one name, a thousand years later another name appears (and names change too, according to language), yet they may be direct descendants of the first people and similar direct descendants may also be found 2,000 miles distant. The second people may also have resulted from tribal intermarriages, or the original people may have been swept away by war and invasions.[24] This happened many times in Europe, Asia and the Far East. Exactly the same theory may be applied to strains of horses. Of course, language may lead the ethnologist to make assumptions as to race and descent of peoples which may or may not be correct.

Since language does not apply to horses, one can only try to fit the jigsaw of the descent of a strain together by trying to discover what bits, etc., were used at different periods and places, the fossils and bones excavated, and whether a given people were equestrian by long usage, had it forced on them or acquired the knowledge through trade or travel east–west, west–east.

[24] For example, since the last war, the entire mobile population of East Prussia moved as refugees a thousand miles westwards, some went even to Canada and South America. The country was occupied by peoples of the Polish nation of another language, dialects and origin. I was driving with a Polish friend in an isolated rural district near Alt Christburg when we lost our way. An aged peasant woman was walking along the road and I said I would ask her the way. "No good," said the Pole in English, "she's a Kaschubin and she won't understand." But the Kaschubin, although autonomous, understood *German* perfectly and I understood her dialect which was completely unintelligible to the Pole, whose home is now in the province.

The autonomous Kaschubin may breed horses, the newly arrived Poles may and do breed horses, the two breeds have different names but they are bred on exactly the same lines, they are not different strains or breeds. The East Prussian horse is bred now a thousand miles west in West Germany. The same horse is bred by the Poles in the state studs in Masuria and they were called Masurian—or more recently Wielkopolska: half-bred Polish. Thus the breed remains essentially the same for maybe generations; or through climatic changes or selective breeding (introductions of stallions from other breeds) or the whim of an official, it may improve or degenerate.

The nomad horse herdsmen, the Ugriers, third millennium B.C., belong to the Mongolian language group, of which both the Finns and the Hungarians are splinter groups. The Finns have never been an equestrian people whilst the Hungarians have produced some of the finest horse-breeders, horsemen and horses in the world. It is enormously important to appreciate this.

It may be noted here that many Asiatic invasions and wars on neighbouring countries, peoples and tribes were basically invasions of, defence of, or re-taking of grazing grounds and pastures. No mercy was shown to prisoners who were either horribly killed or taken as slaves whilst the horses were generally improved by selective breeding and the herds of cattle increased the wealth of the victor.

Sardur III, *c.* 765–733 B.C., took as booty 6,070 horses and 78,197 head of cattle as a result of his seven-year war expeditions in Transcaucasus.

Graves containing horses richly caparisoned with gold ornaments are not uncommon throughout the decades before Christ and this alone would be an indication of the high regard in which horses were held.[25]

On writing this text on Armenia and Urartu in particular, it seems to me very probable that, excluding the recent improvement of type, the Karabakh—in Persian the name means 'black garden'—breed of horses most nearly resembles the old strain that was so highly valued in the first millennium B.C. Now the Karabakh, no longer bred in his original native land nor even of the quality, apparently, of 100 years ago, is used for racing. At the period we are considering, this horse was possibly descended from those earlier horses used by the Hittites, Armenians and Chaldeans who had cultural connections westwards to Greece, the Ukraine and Hungary, and southwards to Egypt.

From archaeological discoveries, it seems that the Chaldeans possessed two strains of horses. One was a small native mountain pony (12 h.h., 125 cm) with a heavy jaw and small feet, and a larger strain which was thought to have been imported into Urartu from the Near East. And the Chaldeans appear on the historical scene at the same time as the Medes and Persians whose earliest settlement is recognized as having been around the Lake Urmia locality. Thus they were neighbours—and this important fact bears upon the subsequent breeding of the far-famed Nisaean breed of horses, and in all probability on the equally famed and similarly built horses of Ferghana which the Chinese so greatly valued.

F. W. König recognizes the township of Arsuniuini, together with the boundary stele which relates to 3,551 riding horses which were kept there. There is similar mention of 2,651 riding horses in another text, which were kept on the southern or northern shore of Lake Van. Everywhere where there was pasture or the possibility of raising horses, the place received the name Arsuniuini. Thus as early as 900 B.C. we have a very clear picture that the Van area was one of great importance for horse breeding. The value of this discovery is increased, as will be seen when we come to the chapter on Thoroughbred horses. This importance is further stressed with the information which the Assyrian king Salmanassar (856 B.C.) gives about the men of Gilzan in the north-west area of

[25] Where are the graves of horses of the twentieth century?

Lake Urmia, bringing tributes of "silver, gold, copper bowls, wood and especially valuable well-bred horses and one and two-humped camels" [Hančar].

King Sargon (721–705 B.C.) also referred to horses which he found in Subi, a country through which his campaign route against Urartu lay, whose inhabitants "were not equalled throughout the whole of Urartu when it came to knowledge [preparation and breeding] of riding horses" [Hančar].

F. W. König found in the rules on sacrifice the existence of a "spirit to protect foals" and a "bridled horse god"—real horse gods in fact. He reaches the conclusion that horse breeding in Urartu had reached a remarkably high standard, since there were horses for the lord's service and horses for working service, and horses were important in forming the basis for building up the army—cavalry horses. It was from the Chaldeans, apparently, that the Churrites or Hurrians, who employed charioteers armed with the bow, were influenced by the usefulness of mounted warriors who were at this stage, however, not yet armed with the lance, which became the favourite weapon of latter-day knights.

It is not appropriate to the contents of this book to go into the various aspects of horse breeding within the complex question of equine sacrificial and cult practices[26] of Indo-European culture nor into the Greek myths and legends[27] which were coupled with cult practices and beliefs. Such beliefs in a variety of forms encompassed Europe, Asia and China over many centuries. All the peoples of the ancient world were involved in the cult of the horse. Its position was godlike, as messenger to the gods when it was sacrificed by the supplicants, and as a source of life and strength when it was eaten at the sacrificial feast. As already mentioned the Germanen kept their special horses in holy reserves, the animals were regarded as almost divine, and no common activity was allowed them.

Bactria had a leaping white horse as its symbol, whilst Hanover, in modern times, has a white horse rampant as the symbol of the province. This covers a span of more than 2,000 years and the distance from what is now Afghanistan to central Germany. Horses, especially white horses, as we have seen in Assur, were bred for these purposes but it is doubtful if they constituted a selected breed or strain and there is no indication whatever that they had any special influence on any type.

The nearest we come to any such interest in the horse in modern times is the American admiration for palomino (golden) horses for parades and similar festivities. Decked out in their silver harness and with the magnificent costumes of the riders, the horses taking part in such parades

[26] See Wilhelm Koppers, *Pferdeöpfer u. Pferdekult der Indo-Germanen*, Pustet Verlag, Salzburg, 1936.

[27] See F. Schachermeyr, *Poseidon u. die Entstehung des griechischen Götterglaubens*, Munich, 1950.

would have been a fitting match for those horses which took place in the cult celebrations of so many hundreds of years ago.

Köppen suggests that some philologists think that Herodotus's white horses were a herd kept especially for cult purposes to which these horses were dedicated. Certainly in Assur this was the case. Wrangel also sees in the horses of the Achasen descendants of the white cult horse. But Köppen goes on to suggest that there is no need to place so much value upon Herodotus's words and says that the Greek word *leukos* does not mean 'white' but 'light', and thus the mouse-grey or dun Tarpan could be meant. Of course, the South Russian steppe Tarpan might have lived in the wild state around the River Hypanis. Another informed writer, Anuczin, suggests that *leukos* could equally mean the yellow/red dun of the Przevalsky horse of Dsungaria and of the onager. Since Herodotus's remarks were applied particularly to the Hypanis area, Anuczin's observations are not really to the point. Later translations of the fifteenth century give the wild horses as *Equi silvestres*, the same Latin description which the chronicler Matthias used at the beginning of the sixteenth century for the wild forest horses of Poland and Lithuania. Herodotus said that the River Bug (Hypanis) rose in a great lake (the area where these wild white horses were supposed to inhabit) and many earlier cartographers have been at great pains to localize the lake which no longer exists. Wapowski, father of Polish cartographers in the fourteenth century, has, however, pointed to the Amadoca Palus, the so-called Amadocian marshes which were said to be shaped like a great lake. Later Gastaldi (1562) and Pograbius (1570) reduced this great marsh to a normal lake and finally it was shown as the source of two rivers, the Słucz and Zbrucz. Buczek observes that Plotomäus's (middle of the second century B.C.) legendary Amadocian swamps are possibly reminiscent of the Polesie swamp, and his idea is that the great swamp in which Herodotus laid the source of the Pontic rivers of Wolhynien and Podolien and the River Boh, which both Herodotus and Wapowski take into account, is probably also reminiscent of the selfsame Polesie swamp.

On these grounds, Vetulani suggests that Herodotus's enigmatic "white horses grazing on the grasslands around a great lake from which the River Boh [Hypanis] springs, in the land of the Sythians", were, in fact, in Poland and indeed around the Polesie swamp, which borders the primeval Bialowiecz forest. During the spring thaws, the swamp does resemble an extended lake. It seems, therefore, likely that these are the same wild horses which Matthias and others mention, the same *Equi sylvestres* which were living in the forests of Poland and Lithuania during the fifteenth century.

In fact, the Hypanis does not appear to have its source exactly in the Pripet marshes, as does the larger river the Borysthenes (River Dnieper). Poles'ye is a swampy lowland which includes the Pripet basin and the adjacent part of the Dnieper lowland, roughly an area comprising 60,000 square miles, is partly covered with pine and oak forests. The province of Bialystok in which lies the forest of Bialowiecz or Bialystok is low-lying and is watered by the Rivers Bug and Narew.

From his analysis, Vetulani assumes that the white horses were Tarpan, *Equus przevalskii gmelini* Antonius, in their white winter coats, just as today the regenerated Tarpan in the reserves carries a white winter coat, i.e. without pigmentation. From other sources one learns that a similar coat-colour change—from dun or mouse-grey—takes place amongst the ponies of the peasants living in the neighbourhood of the primeval Bialowiecz forest. Vetulani was writing in 1940. When I visited the descendants of these same regenerated Bialowiecz Tarpan in the reserve at Popiellen in East Prussia, a primeval forest also bordering a great lake, the Sperding Lake, those in charge were unable to say if the coat did change to white, although it does definitely become lighter on many of the Tarpan there. What was especially curious were the stripes on the stallion.

A number of northern animals and birds have permanently white coats. Amongst these are numbered the polar bear (*Ursus maritimus* Phipps), the snowy owl (*Nctea myctea* L.), and the northern falcon (*Hierofalco Articus* Holb). Others change their coats to white in winter. These are the arctic fox (*Canis (Alopex) lagopus* L.), ermine (*Mustela (Mustela) erminea* L.), arctic hare (*Lepus varriouis* Mill), lemming (*Dicrostonyx (Myodes) torquatus* Pall), ptarmigan (*Lagopus lagopus* (L) Br.), polar wolf (*Canis tundrarum* Mill), and the reindeer (*Rangifer tarandus* L.), the forest Tarpan (*Equus przevalskii gmelini sylvaticus*), the now extinct Lofoten Islands pony, and the Siberian Yakut ponies who also belong to the Tarpan group; although we do not know if some Yakut ponies retain the white colour throughout the year. Hutten-Czapski declares that the Steppe Tarpan, who also had a mouse-grey or dun colour during spring and summer, grew long, white top hairs of varying lengths as a protection against the winter weather, and these hairs were shed again in May. This procedure is normal amongst many of the Konik group of Eastern Europe.

Vetulani indicates that the forest Tarpan (*Equus sylvestris*) is a secondary variable forest form of the original steppe Tarpan and that the change of colour is due to the changed climatic conditions and especially to the lack of daylight during the central and northern winter.

> That the climate does undoubtedly have an effect on the pigmentation of coat colour, was observed during the mild winter of 1938–39 when the Stallion 'Tref' changed his coat to a lesser degree of white. The coat was not snow-white but dirty white in places. The mare 'Borowka' showed amongst her winter white coat on her quarters also pale pink hairs e.g. roan. Probably the ability to change the coat amongst the Koniks of the Bialowiecz forest

was due to the breeding of typical pairs so that an atavistic characteristic inherited from the forebears came to the fore.

Thirty years after Vetulani studied the white factor of the forest Tarpan, I was able to photograph a striped Tarpan stallion. This horse had a brown coat with a strongly marked dorsal stripe and stripes over his neck, withers, back and quarters. There were at the time (1964) four such striped animals in the reserve. One in particular had very definite stripes on the face, particularly the forehead. Again we must assume a strong atavistic characteristic. Further primitive characteristics which deserve comment are the fact that these regenerated Tarpan never abort, never catch cold and if they do suffer wound injuries, these heal very quickly without outside help. Western European breeders have never given sufficient thought to the value of the work that was being done at Popiellen (in 1964). Scientists there thought that this work could have proved of the greatest help to breeders of the Thoroughbred horse where the abortion rate is fairly high and the tendency to catch cold—equine influenza—appalling.

4. The Horses of the Peoples of Antiquity

As we have seen, to the peoples of Akkad, Sumer and Elam in the third millennium B.C., the horse was, if known of at all, a rare animal. They did not even use the word 'horse' but an equivalent to 'equid'. And when later 'donkey of the mountains' was referred to, it is thought that *horse* was meant. The domesticated horse may have made an appearance in Mesopotamia as early as 2300 B.C. But it was many decades later before the horse was mentioned on cuneiform tablets.

In the Song of King Šulgi, *c.* 2000 B.C., the 'foreign ass' is mentioned. Šulgi likens himself to the noble horse:

> A *proud* ass [literally, ass with long legs—mule?] hastens along the road
> A fast horse with waving tail,
> A stallion donkey of Sakans, who knows how to run, am I.

But this horse of King Šulgi was already so well-trained that it covered the distance from Ur to Nibru in "15 double hours": 150 kilometres or 75 miles. So one can see that the Babylonians or the man in charge of King Šulgi's horses knew quite a lot about horses even at an early period, and although 'horse' in the singular is mentioned, since a light chariot would have been used, this horse would have been one of a pair.

In the third millennium, to the north of Akkad in Syria, Asia Minor and Armenia, there were as yet only native city states or small principalities and there are no records of wild horses except in the area that was later to be Armenia from the Black Sea to the Caspian Sea and the Caucasus Mountains. To the north, the discoveries at Maikop in Kuban on the eastern shores of the Black Sea show that horse breeding was in process. In fact, it appears that it was the Przevalsky horse and not the Tarpan as one might have expected.

As O. R. Gurney writes, history begins in Anatolia with the arrival of Assyrian traders in 1900 B.C. They used the cuneiform script of Babylonia and many clay tablets recording their deals have been found. The land of Hatti in Anatolia was inhabited by non-Indo-European Hattians and it was settled by Indo-European invaders who subsequently called themselves Hittites. We shall return to them later.

Let us now briefly examine how it would be possible to develop the horse's usefulness. As I have observed elsewhere, in forested areas riding is more likely than driving, and the type of horse available, the heavier European horse, is more suitable to carry weight than the slighter pony of the desert. The method of control, too, can be exercised far more effectively from the *back of a horse* than from the length of the reins when driving. The yoke, which fits an ox very well, is not a very effective piece of harness for horses to wear. Ox-carts were cumbersome for a lively pair of ponies to pull. Therefore the early Mesopotamians who realized the usefulness of the horse—or perhaps its possibilities—had to do a considerable amount of inventing.

We should by now be able to appreciate that:

1. The horse was not common and the oriental strain, when it was introduced, came from the north and the east. Later a heavier strain was introduced from the west.

2. This oriental horse was of light conformation and its temperament was quite different from that of the ox, ass (which had been imported from Africa) or the native onager. It was also a different animal from the mule.

3. It was much faster and therefore a new kind of harness had to be invented. There is extant text concerning "horses bound together with the yoke".

4. To make full use of the horse's mobility, a lighter vehicle with spoked wheels had to be designed.

5. Places where these animals could be bred had to be organized and now horses must be imported until these areas were self-sufficient.

There are frequent texts which refer to tributes, imports, presents, and also requisitions, all having the opportunity of introducing fresh blood.

Bits of bone, leather or hemp—later iron, bronze, silver and even gold—have been found. Copper was very rarely used since these ancient peoples were clearly aware that oxidization could easily injure a horse's mouth. Since many of the early bronze bits were twisted (snaffles), after the manner of hemp, it is likely that horses were bitted earlier than we suppose. Both leather and hemp would disintegrate with time and thus could not be found.

Now, in the second millennium B.C., the lands of the Euphrates–Tigris basin, Syria and Anatolia and east of Anatolia towards the Caspian Sea were to be settled by adventurous newcomers. And it is not entirely out of the question to suppose that some Indo-Europeans had come from the west or had western connexions. The Tripolje civilization, as already shown in the last chapter, had precedence in the knowledge of breeding and using horses. Pia Laviosa-Zambotti writes: "At this epoch the most archaic Indo-European races [Hittites, Achaeans, proto-Latins] established themselves along the banks of the Danube and, at a later date, invaded the Mediterranean peninsulas [Anatolia, the Balkans, Italy]."[1]

[1] Larousse *Encyclopaedia of Prehistoric and Ancient Art.*

On the eastern border of the kingdom of Hatti now occupied by the Hittites, there appeared from Armenia eastwards a people called the Churri or Churrians (Hurrians). They settled around Lake Van and were to become the greatest expansionist power of the second millennium. They were, too, largely responsible for the later Hyksos movement into Egypt (1670–1570 B.C.). The Churri language was neither Indo-European nor Semitic. These people had a considerable knowledge of horses and there are many references to brown horses. This energetic people founded the state of Mitanni and their political power spread from Lake Van to Aššur, from the coast of Palestine to the Zagros Mountains. The Churri themselves were governed by an Indo-Iranian ruling class called the Marijanni or Marjannu (Chinese *ma* = horse) meaning chariot warriors. Probably the Marijanni were the earliest knightly order or class of warrior nobility. Their western neighbours, the Hittites, were also enthusiastic horsemen and particularly skilled in making harness. There are archaeological pointers to the probability that they conquered parts of their new country as early as the second millennium B.C. and remained in comparative anonymity for several centuries.

But this period was one of considerable unrest and displacement of peoples caused by the desire for political power amongst the ruling castes. The horse was becoming the new weapon of war, excursions into neighbouring states and rapid victory belonged to the ruler with the most mobile army. The period can be likened to the expansion of states during the industrial age or the expansion of political ideologies in the nuclear age.

Hančar gives details of a letter—one of the few cuneiform inscriptions from the eighteenth to seventeenth centuries B.C.—which mentions the horse. It is an order from King Samši-Adad of Aššur, who was a contemporary and opponent of Hammurabi, the distinguished ruler of the First Dynasty of Babylonia in the eighteenth century B.C. Samši-Adad writes to his son Jasmach-Adad and asks him to send to Aššur pairs of horses together with ceremonial racing war chariots from Til-sah-Annim (Tschagar-Bazar on the Chabur in northern Syria) for the processions and celebrations for New Year. And Syrian horses are meant when we read: "The price of these horses of ours in Qatna are 600 (Sekel) silver." To confirm that they are as yet not so common, someone is instructed: "Take a measure of corn as food for the horses, so that the horses may eat and may not be hungry."

Another famous inscription relating to horses of the same period is a reply from Aplachanda, the King of Karkemiš, a state on the Upper Euphrates, to the envoy of Zimri-Lim, King of Mari, also a contemporary of Hammurabi. "I spoke to Aplachanda about the white horses and he replied: 'White horses for the war-chariots are not available. I will send for them where they are to be had, white horses shall be driven hither. Until then I will have bay horses brought from Charsamna'."

The Bogharzköi Hittite text gives the following forms of the name: Harsamna—Harsumna—Har Sumnasa. And Weidner says that "A thousand years after Zimri-Lim, Harsamna was still known as the *Mountains of the horses*" and must have been a well-known stud, or breeding centre for many centuries. Probably it was situated in modern south-east Anatolia and was in existence long before the Indo-European period. According to Cornelius, Harsamna lies near Herik. According to Balkans, the more probable locality is in the Uzun Yayla district, which, says Professor Kammenhuber, is today still a recognized district for horse breeding. It is east of Kayseri and south of Sivas. There is a road through the mountains which connects both places.

Cilicia was also a noted country for supplying horses to other states. The great Hittite King Mursili II (1345–1315 B.C.) had an agreement with the King of Kizwadna (Cilicia) to supply 100 bridled horses and 1,000 men in time of war, that is one war chariot per 20 foot-soldiers. And by 'bridled', they are meant to be trained horses for war.

The town of Karkemiš was a great horse market even in Hammurabi's time. We do not know whether it was solely the king's privilege or whether it included a much larger circle of interested dealers. But as Weidner says, the letter of Aplachanda is interesting on two points. First, that Zimri-Lim wanted white horses for his war chariot and that there were studs outside Karkemiš and not including Charsamna, where horses were bred apparently for purposes of war. It is tantalizing that there is no further information as to the position of this stud where these special white horses were to be found.

There is yet a further letter from Babylonia in which the addressee is asked by the sender to send him two white horses. This letter belongs to the reign of Samsuilunas.

It was nearly a thousand years after the first established appearance of the domesticated horse that archives of clay tablets were being built up in the fifteenth century B.C. The great library of Bogharzköi was possibly commenced by King Telepinu; the archives of Mari on the Euphrates, of Tell Halaf on the River Chabur, of Nuzi near Aššur and Susa in Elam all tell of the historical progress that was being made by the horse. This one animal had become of the greatest importance and played the major role in the political development of all these states and of Egypt as well.

One of the most important discoveries amongst the cuneiform tablets from Bogharzköi were those dealing with the training of war chariot-horses. It is the first 'book' ever to be written on training horses. There were several copies, some produced at a later date, but the original was composed by a man called Kikkuli who with his co-workers came from Mitanni. The Kikkuli text gives the exactest details of the Aryans' sound method of training horses. Their language had, it seems, already become a dead language although their experience under the Hurrians (Mitanni) lived on. In fact, it would appear that the Hittites had already, at this time, overthrown the state of Mitanni with their own war chariots before Kikkuli and his colleagues wrote their treatise some time after 1350 B.C. And none the less the influence of Mitanni remained very strong, although it is possible that these Mitanni horse-masters were emigrants or refugees!

Kikkuli's method of training covered exercise by day and night and included trotting and galloping over various distances, swimming, sweating, hot and cold bathing, and racing which lasted over several days and nights—a fairly exhausting sort of endurance test. The period of training covered 184 days.

Here it is appropriate to note that a third training text has come to light more recently, which shows the writer to be clearly a Hittite. He treats his horses much more kindly and grooming and feeding play an important role, as indeed they should. Also the distances are considerably reduced. In the one instance too much was required and in the latter, perhaps too little.

Horses were fed as ordered on the following fodder: hay, made as we make it today—not baled, of course; barley—or perhaps corn (wheat) is more generally meant, because barley is not too good for the wind; oats are not grown in Anatolia and the Hittite horses in training would not have received it; straw chaff, which today is not used as much as it could be. An ointment or oil was also rubbed into the horses' skins.

We may wonder how it was that the rulers of Mitanni had this specialized knowledge of horses. In the first instance it seems that they themselves had come from the east. They worshipped the Indo-Aryan deities such as Varuna and Indra and the Nasatya twins. And their personal names, too, had a similar origin. It seems therefore that this Aryan people brought with them their special knowledge of breeding and training horses. Also they must have brought their horses with them, probably *Equus orientalis*, the wiry little horse of the desert and steppe.

One result of the specialized training of the Hittite chariot horses was the victory at the Battle of Kadesh, 1286 B.C. The King of Kadesh, possibly a vassal of the Egyptian Pharaoh whose influence had reached Syria, came out to do battle and was overwhelmed by Hittite chariotry under the leadership of King Suppilulumas. This was one of the most famous battles in ancient history. We cannot here go into further detail concerning the historical events in the Near East since, although horses undoubtedly affected the turn of events, the events only affected the horses in so far that greater numbers were bred.

King Burnaburias II of Babylonia (1367–1346 B.C.) received over 600 kilogrammes of gold from Amenophis IV of Egypt and Burnaburias adds in his letter to the Egyptian Pharaoh: "They say that the roads are bad, water cut off and weather warm so I have not sent you many presents. I have only sent 4 *Minen* [2 kilogrammes] of lapis lazuli as the main present to my brother and five pairs of horses. As soon as the weather is better, I shall give my envoy many lovely presents for my brother."

There is, further, a clay tablet from Assyria which Aššur-uballit I sent to the same Pharaoh: "I send you two white horses from my stables."

Later there are frequent references in Assyrian texts to white horses. One in particular is directed to law-breakers who have to "tie two white horses to the feet of the god Aššur", or, "four white horses are to be bound to the feet of the moon god". Other wrong-doers had to keep the temple supplied with white horses and therefore there must have been quite a number within the temple precincts at that time. It also suggests that white horses must have been available locally within the kingdom and so perhaps they came from one of the studs listed at the end of this chapter.

We see, then, that the Assyrians favoured white horses especially for their religious cults and processions and kept a number to draw the holy chariot of the god Aššur.[2]

Although white horses were so popular amongst the Assyrians, especially for ceremonial purposes, there is no account of them in neighbouring Babylonia. The Indo-Europeans, too, favoured white horses because they saw in these beautiful pure white animals the familiars of the gods, whilst the Germanen of a later date kept white horses in holy reserves and they were not allowed to do common work. (The coat-of-arms of Hannover is a white horse rampant[3]).

The Persians especially liked white horses and eight were harnessed to Zeus's holy chariot at the time of Xerxes. There are many more examples even during Roman times.

Weidner suggests the conclusion that the Indo-Europeans therefore had a cultural influence on the Assyrians and later the Persians. The former were subordinate for a long time to the east Indo-European Indians or Indo-Iranians, in their Mitanni kings. The kingdom of Mitanni bordered the Hittite kingdom and as early as the second millennium Mitanni knights had established small principalities in Syria and southwards in Palestine.

Tusratta of Mitanni wrote, in the first of thirteen letters to Amenophis III and IV, after greetings and good wishes for his horses and his chariots, of his successes both domestic and against the Chatti land (Hatti): "See here, I have sent you a chariot, two horses, a youth and a maiden as booty from Chatti. As a present for my brother I have sent you five pairs of horses."

[2] This reminds me that the Carthusian Order in Spain and Great Britain, in the Middle Ages, favoured white horses and white cattle.

[3] See page 175 on the royal cream horses.

Between 1630 and 1530 B.C. hordes of Cassites appeared in Babylonia. Their language connects them with Lauristan, the Caucasus and the Caspian Sea area. In domestic documents (cuneiform tablets) these people are mentioned as early as the seventeenth century B.C., as harvest workers and peasants. So it would seem that they too did not arrive with warlike intentions, although after Hammurabi's death they threatened Babylonia, were warded off by his son and finally settled in the north. They were horse breeders and exceptional horsemen. Horses were honoured almost to a cult. Their language includes the word 'stallion' but apparently has no word for 'mare'. The colours of their horses are interesting and are deciphered from logograms:[4]

Mi = black
Sa_5 = red or bay
Babbar = white, grey
Gun_3 = odd-coloured, skewbald
Sirpi = brown
Minzir = chestnut
Pirmah, *Pirmuh*, *Purmah* = partly-coloured like the Turkish *sekili* "with white spots on the fetlocks"
Laga (s) tak (k) as = shades of bay
Dakas = star, white spots, perhaps dappled
Hulalam = blaze, or possibly a Cassite word taken from Akkad, *hulalu*, meaning "a jewel probably of a yellow colour"

There are many other words for colours, but the meanings are not known to German scholars. It also indicates that the horses were advanced in selective breeding, for they were no longer of primitive colours like the Asiatic wild horse's colours of shades of chestnut dun or the Tarpan which is brown, mouse-dun.

The Cassites attached great value to the pedigree and considered the sire to be the dominant factor. Ordinary horses had no names and generally only the stallions would be given names. Many of the names thus given refer to faults, colours or animal names. Some are Cassite and some Akkad. Many were called after their colour, and their breeding or pedigree would be given, like "Descendant of the Bay".

Another word is *tur*, foal. The foals ran alongside their dams when the mares were in harness, or perhaps were tied to their dams to accustom them to harness. This ancient practice still survives in eastern Europe.

Sumaktar = half-bred, meaning sire unknown.

All the difficulties from which we may still suffer today when dealing with insufficiently trained horses were known more than 3,500 years ago to Cassite or Akkad horsemen. For example, "When a prince drives in a chariot and the horses keep jumping about. . . . When the horse har-

[4] A. Salonen, *Hippologica Accadica*.

nessed to the chariot of the god becomes unmanageable . . . becomes unmanageable so that the pole breaks . . . becomes unmanageable and gets frightened . . . becomes unmanageable and throws himself on to a person." Since the reference is to the chariot of the god, it means that this was a procession and we know only too well that such behaviour on the part of a horse can ruin a procession.

Accidents happened then, as today. "The king [or the prince] mounted the two-wheel chariot and fell on his right side and the wheel on the right side or the horse on the right side smashed his head."

There are references to horses at pasture, horses used to war, and horses bound together with the yoke. There are also two charming little stories which have come down to us over 4,000 years. "A rearing horse, on covering a female mule [ass?], whispered in her ear: 'The foal that thou shalt have, shall be a runner like me; you shall not make a donkey that carries panniers'."

And the fable of the horse and the ox:

> The horse says: "I am dressed [harnessed], like a robe it shows off my feet, without me the mare's foal would not be able to pull kings, princes, nobles and gentlemen on their way." The ox observes that the harness is made out of his skin.
>
> But the horse shows off with his voice and his fast action and he further boasts that he gets the best fodder: "They choose for me the finest herbs and the green of the earth. For the noble horse they prepare his drink." But then Anu [the donkey?] comes and says: "Horse, your office is difficult. You and the riding ass, you carry the panniers, like the caravan road you bring the four winds nearer."

Then it seems that the horse rudely replied.

> ". . . spoke to the pack donkey: 'you're not big . . . prick and whip are your lot, you look like a box . . . really you will never carry an envoy.' When the pack donkey heard this, he screamed angrily . . . made a loud noise." [Salonen]

The great value set upon the horse may be seen from the formal conclusion to letters sent by Cassite and Akkad kings to their brothers the Pharaohs of Egypt: "It shall be well with you, your country, your house, your wives, your belongings, your horses and your chariots." And King Chatusili III observes that in Babylonia there are more horses than straw. It may have been a diplomatic expression or hint that he would appreciate a few surplus horses!

There are handed down to us Akkadian words for bright chestnut, *samu*; black, *salmu*; and grey, *pesu*. There are remarks as to 'horses from herds', 'rations for horses', and 'corn'. And the informative letter of King Hattusili II to Kadas-man-Enlil of Babylonia includes the following:

> So I ask my brother. Send me horses, well-grown colts. The stallions which your father and those which you sent were good, but have become old nags;

32. A bronze relief of a campaign in north Syria, 858 B.C. It shows Assyrian archers and chariots at the storming of Dabign, and female captives, mules and dromedaries on the march

33. A campaign in Armenia, 857 B.C., showing the tribute of horses from the men of Gilzan

34. King Assurbanipal on a lion hunt, seventh century B.C. His capital was in a horse-breeding area

35. The horse of Ferghana. Excavated from a tomb of the T'ang Dynasty at Astana Turfan

> and old horses [are no good?]. In Hatti it is very cold and old horses do not live long. Therefore send me, my brother, young colts. There are enough nags in my country.

(Kammenhuber states that this letter was sent by Hattusili III and was dated several centuries later than the generally accepted date.)

In Hatti there were plenty of horses but they were of poor quality, being small and not particularly suitable for galloping. Able horses had to be imported. The main condition is the necessity that acclimatization takes place. The sort of horses required were not bred in Hatti and only young horses would do since they could be hardened off as they came from a warmer climate.

Horses played a very large role in political and diplomatic dealings. The kings of Babylonia, Assyria, Hatti and the princes of Syria and Palestine sent not only their daughters but horses as well to the Egyptian Pharaohs and begged for gold in exchange!

This great expansion of the horse population took place within a matter of a few hundred years in a part of the world where they were not primarily native, although I suspect that the small, poor quality horses referred to as being present in Hatti could mean that these were the descendants of wild horses. However, they were brought in from the north and east and also probably from the west, as we shall see.

Again in a matter of a century or two, studs and areas where horses were specifically bred became well known. These areas were much too large for our present conception of a stud,[5] but indeed they were studs in the widest interpretation of the meaning.

The old Slavonic word *stado* means 'herd of horses' and combines with the old German (both belong to the Indo-European language group) *stuota*, mare.[6] Literally thousands of horses were raised at these studs, generally on the *tabun*, or herd, principle, from Cilicia in southern Anatolia to Ferghana and the district on the River Ili—a distance of at least 3,000 miles as the crow flies—to the east, whence the Chinese rulers sent for horses.

In the Middle East, the most important perhaps were in Elam and Syria, in north Mesopotamia on the southern slopes of the mountains and especially in the Lake Van area which was inhabited by the Churrians. Three thousand years later, horses from this very area were imported into Great Britain and helped to lay the foundations for the most important breed of horses in the world—the Thoroughbred horse.

Most of these areas bred the small desert and steppe horse, *Equus orientalis*, who was perhaps even the ancestor of the Arab. Some are named as breeds.

[5] See Empress Maria Theresa, *The Founding of the Hungarian State Stud (Mesöhegyes)*.

[6] Modern German *Stute*, also old German *Marah, Meriha*; *Mähre, Mare*.

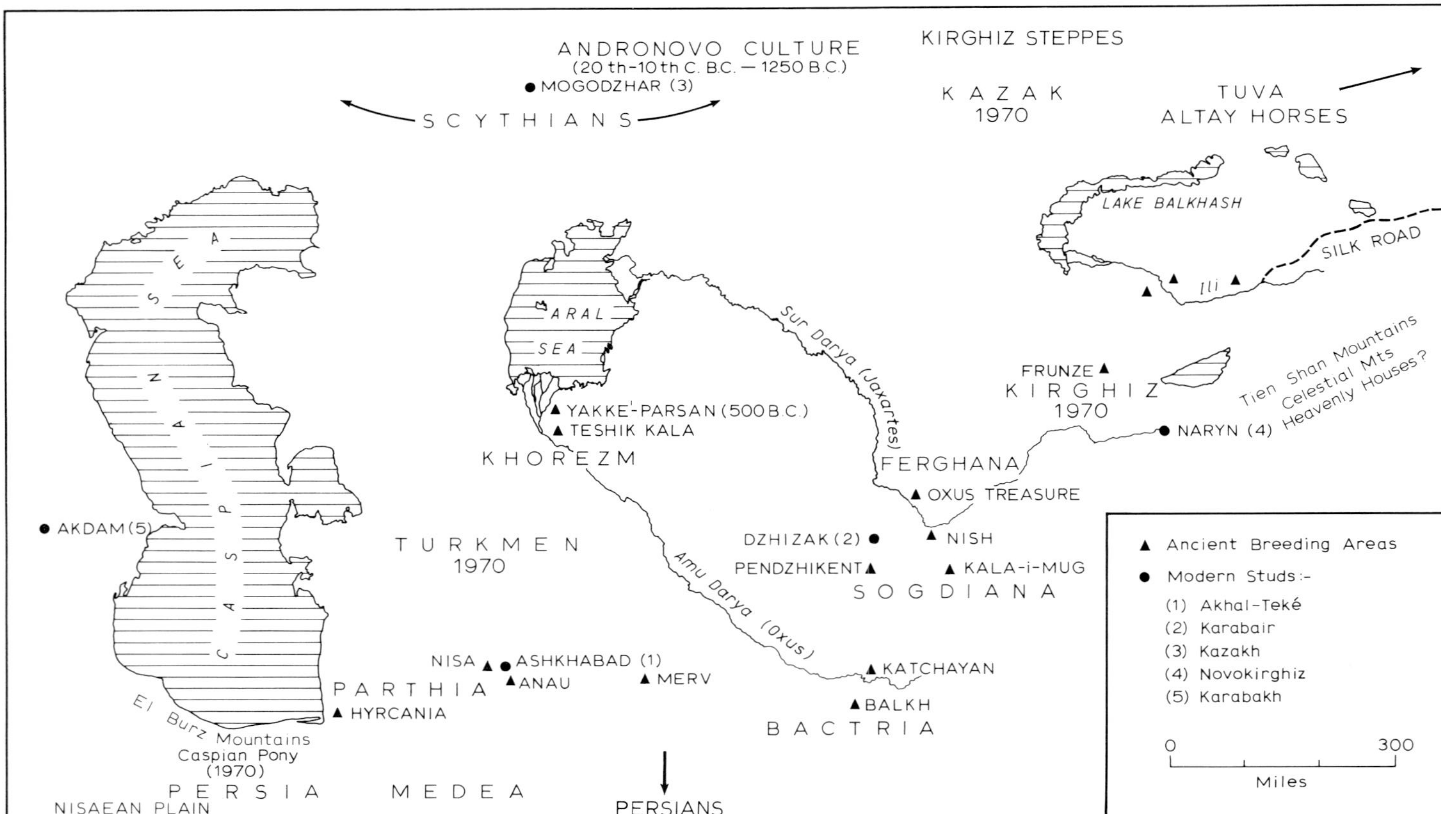

36. Map: Breeding areas of Central Asia

1. In Persia, Elam = the towns or districts of Kusa, Mesa and Parsua bred Susian horses.

2. In Assyria and northern Mesopotamia = Aššur, Kalhu, Arrapha and Lahiru, Barhalza and Guzana which was famous both for its horses and mules in the time of King Sargon.

3. Urmia and the Van district: Gilzan and Assurna. Sargon recommends new blood from the Gilzan land.[7]

4. Syria—Palestine. There are records of tributes from:
Qar-m-e = *Da-a-na* = *Kul-la-m-a* = *Ar-pad-a* = *I-sa-na*.
Salonen suggests that Qarme is probably the biblical Qarnaim, eastwards from Jordan; Kullamia the biblical Kalneh, in northern Syria. Isana was probably on the River Zob and Da'ana near Aleppo. There was also *Man-zu-at-e* in northern Syria between Karkar and Aleppo.

5. Asia Minor. *Me-lit-a-a* = Mehdai of Tabal, the latterday Roman town and province Melitene.

6. Arabia = Musri: Misri. Musri lay in the basin of the Nile eastwards and is now Egypt and this is the only reference which I can find which refers either to Arabia or to horses therefrom *at this period*. Weidner translates a text: "Silkanni or Silḫeni, the King of Misri, brought twelve large horses from the country Misri, the sort which are not to be found in my country." Silkanni is a contemporary of Sargon II, King of Assyria, 722–705 B.C.

Weidner, however, insists that Misri is not Egypt but Arabia, north Arabia, south of Palestine on the land narrows to Egypt. He says that horses would more likely belong to Arabia than to Egypt where, in any case, they were imported from Asia Minor and Syria. Even so Misri lay in the path of such importations. Also, since they are especially mentioned as large horses, one wonders if Misri had, through Phoenician shipping channels, a route to Greece and Mykenai (Peloponnisos) where the remains of large horses have also been discovered, as well as those of a smaller strain. Here is possibly a connection with *Equus robustus* of Europe? Misri horses were used as chariot horses. Tributes of horses were delivered monthly.

Now in the ninth century B.C. there came into being, partly overlapping the former state of Mitanni and partly occupying lands to the east, a state called Urartu. Very likely the inhabitants were descendants of the Churrians. They raised horses for the army of Sargon, "saddle horses which are not to be equalled anywhere", and Assarhaddon mentions "big" horses as tribute. He says, "The fiery galloping noble horses, the team in my two-wheel war-chariot, sank in their blood as if in a river." The Urartans certainly were amongst the foremost horse breeders. Sargon I found the horses around Lake Urmia to be so numerous that they could not be ridden (tamed?). Whilst Rusa of

[7] See Corduène and horses of the Gordeen breed.

Urartu boasted: "With my two horses and my driver, I conquered the kingdom of Urartu with my own hands."

Jacobsen writes that Sargon I makes his thoroughbreds thrive, makes his teams perfect. And a century later Assurbanipal said: "I mounted the noble horses, I rode fiery thoroughbreds." Assyria had used the popular oriental horse, which we can see on many engravings and stone carvings, until approximately the tenth century B.C. when, after a lapse in its fortunes, the country began to recover. This was due to the effectiveness of *mounted* cavalry over the now obsolete chariot. These cavalry horses were bigger and many were supplied by the Medes to the east, from their extensive herds.

During the first and second millennia, there were no made-up roads throughout the Middle East, except in towns, but there were many routes, caravan-ways, foot-ways and cattle paths and the 'royal roads' which were used for military purposes. There were army posts and barracks all along these roads and couriers were stationed as well. These couriers used two-wheel carts and they probably transported people and small loads.

The earliest armour for horses is mentioned in the thirteenth century B.C. and is generally in the Hurrian language, Hittite, and sometimes Cassite. The armour consisted of a leather protection for the horse's forequarters, sometimes made of scales or bronze metal parts. A sort of headguard or helmet was also used. Rugs for horses are also mentioned in ancient texts.

The first mention of riding appears in the Alalakh text of the thirteenth century B.C. The word *pāgumu* is used for saddle but there is no indication as to whether it was used on a donkey, mule or horse. The earliest information on actual riding horses is around the twelfth century B.C. under Nebuchadrezzar I (1130 B.C.) when saddle horses were brought to Babylon. In 860 B.C. Assur-na-sirpal introduced mounted soldiers into his army.

The strain of horses was improving from the first millennium onwards. This was probably due to better feeding which was possible as farming methods improved. The discoveries in Lauristan are marked by the discoveries of the increasing variety of bits. Burials show that the love of horse breeding was uppermost and the dead were often laid to rest with their heads on the bridle. The bits were of two kinds: the straight mouthpiece and rein rings decorated with horses, cattle, goats, etc., which are reminiscent of the second millennium B.C. when Lauristan was occupied by the Cassites. The second type is the simple snaffle with straight cheekpieces which sometimes end in animals' heads. Generally these bits were made specially for burial purposes and were not in everyday use, although it can be assumed that similar bits were part of everyday harness.

Medes, Persians, Cossäer, Bactrians and Caspians, all peoples of north-east Iran, constituted the first cavalry divisions of the Archämed army, whose main weapon was the bow. But generally in Assyria, Syria, Egypt and even India the two-wheel chariot was the chief instrument of war and the symbol of power until the Medes and Persians proved that mounted cavalry were superior, and the peoples of the above countries were obliged to learn the art of riding. They never willingly wore the costume which marked the rider, namely trousers.

There are records that Thracians from Europe were riding *c.* 1200 B.C. However, before we go into further details on the breeding of ninth-century B.C. saddle horses in the Near East, we should examine the evolution of the horse in North Africa and Egypt and the countries which came under the dominance of the Pharaohs.

It is not safe to assume, as so many writers have done, that there were no horses in North Africa prior to the Hyksos rule in Egypt (1670–1570 B.C.) because just the opposite is historically more than probable. At any rate the close connection of Morocco to the Iberian peninsula and the very similar conformation and colour of the primitive Sorraia horse to the more developed Berber horse, make an interchange between the peoples of the countries a probability, even if there is no written evidence to rely on.

Not generally known is the fact that from prehistoric times there has existed in the Iberian peninsula a wild horse with a convex profile. Publius Vegetius[8] in his *Mulomedicina* states definitely that the African horse is of Spanish origin. Also a few sacrificial bronzes of *c.* 300 B.C. (one from Despeñaperos, Province Jaén) show markedly ram-nosed horses and horses with concave faces.

There seems to be strong evidence of Mesolithic domestication of horses in Cantabria since there are extant cave drawings dating from the fifth millennium of horses being led, and from the fourth millennium of horses being ridden. Perhaps these are another indication that riding originated in Europe.

The late Ruy d'Andrade, one of the greatest equestrian historians, produced a map of the peninsula and of Morocco showing shaded areas where primitive horses of antiquity were to be found. In Iberia there were apparently herds or droves of wild horses in most of the principal river valleys and in Morocco, around the basin of the Moulouya and the River Loukos on the west coast.

The similarity of the North African Berber[9] to the Sorraia and the subsequent mingling of both in the South American *Criollo* cannot be denied. All have the pronounced and unmistakable convex face. In this

[8] Flavius Vegetius Renatus, *Digestorum Artis Mulomedicina libri*, A.D. 400, a book on veterinary medicine.

[9] Known commonly and incorrectly as Barb.

37. Relief in Temple of Karnak at Luxor, Egypt

38. A painting depicting a horse and probably mules from the tomb of Mab-Amon, Egypt

horse, as we have already seen, there is an entirely different structure of the skull, in particular of the face profile and the jaw-cheek profile. Whereas the Przevalsky and Tarpan descendants are shown to have a less or more concave face in possibly varying lengths, the Berber horse has this markedly convex face and almost primitive jaw-cheek profile. When crossed with Arab, the convex profile is still dominant but given the right—or wrong—climatic conditions and over thousands of generations the Berber (and Sorraia) probably evolved from *Equus robustus* (cob) and *Equus przevalskii gmelini* (Tarpan).

Without apparently the slightest knowledge of the existance of the Sorraia horse, its age, conformation or locality, Antonius writes: "We can accept without the slightest doubt, that a more or less uniform strain strongly inclined to the ram-nose was spread from the Spanish peninsula to the Atlas mountains, before the better-bred Arab arrived via Libya, the earliest probably at the time of the Egyptian New Kingdom." Antonius further suggests that it was the Celts who brought the Tarpan—but, I think, more likely *Equus europaeus*—Garranos de Minha strain with them from which the bronze horse with the concave face was taken. On the other hand we know from the cave paintings that *both* the convex and concave face strains of horses must have been present in Southern France and Iberia many thousands of years prior to the arrival of the Celts.

Even Marcus Aurelius Olympinus Nemisianus mentions the ugly heads of the African desert horses, since the Romans themselves were used to the concave Tarpan head. Marcus Aurelius calls them *turpe caput*. None the less, the Romans admired the North African horses because of their stamina and kindly dispositions. Both these qualities are also emphasized by other writers.

Berbery *not* Barbary consisted of that coast area of North Africa which was Numidia and Mauritania (hence the later term Moorish horses, the horses of the Moors who were in fact generally Berber). But the land of the Berbers was conquered by the Arabs in the middle of the seventh century A.D. and consequently the horses of the Arabs added a new breed to the native population. Claudius Aelianus[10] said of the Berber horses: "They are small and not very beautiful, but extraordinarily fast and strong and withal so tame that they can be ridden without a bit or reins and can be guided simply by a cane, so that only a lead rope is necessary on the halter."

Although much of the original Berber blood became mixed with Arabian after the invasions of the seventh century A.D., Berber characteristics still prevailed: convex face, high wither, flat belly, sloping quarters with tail set on low, very docile, quiet, strong with plenty of

[10] A Greek writer of the second and third centuries A.D., who wrote *On the Nature of Animals*.

stamina. And the breed still known as Berber achieved such repute that throughout the Middle Ages[11] they were regarded as amongst the best horses in the world and were to be found in the studs and stables of the great in both Africa and Europe.

Marcus Fugger, of whom more later, wrote:

> The Berber horses are wonderful noble horses, they are fast and strong and can stand great hardship, and if they were as big as European horses, there would be no better horses for use in war. The ancients too valued these horses a lot and even now they are highly regarded especially amongst the Turks.[12]

Symachus, a writer of the second century A.D., says that Spanish horses were famed because of their speed, and that they were exported to Syria. Thus the Islamic invasions brought not only new Arab blood to Spain but also the old related blood from North Africa, in what quantities or how mixed will never be known and thus we have the foundations for the later Spanish or Andalusian breed of horses famed throughout Europe.

One hundred and fifty years ago, Ammon wrote that at this time the Berber horses were no longer what they once were, even though in parts of Berbery (Morocco?) excellent horses are still to be found and that there were two strains of horses: "the ordinary horses of the country which are in no way remarkable and look like any other ordinary horse. And then those horses of nobility, which we in Europe call Berber, incorrectly Barb".

However, according to Leo Africanus and Ammon himself, it would appear that the well-bred Berber was often Arabian or largely of Arabian descent because they were still bred by the wandering nomad Bedouin Arabs. After the Arab conquest, the remaining Islam nomads settled in towns and villages and mixed with the original inhabitants and were subsequently called Moors, and the breeding of horses suffered from neglect.

The better-class horses rarely exceeded 14 h.h., had small fine heads with convex profile, fine ears, thin long necks, flat thin shoulders, flat bellies, straight backs, strong short bone, were well-muscled, had long croup with tail well set and carried high. Hair was smooth and soft. The biggest were white or grey, whilst the best-looking were generally dark-brown or chestnut. Their speed, strength and hardiness were equal to that of the desert Arab, and any horse that was fast enough to catch an ostrich was worth 100 camels or 1,000 ducats. Such horses were, not surprisingly, very rare!

Even the commoner horses had their merits because they were also fast, strong, willing, patient and good-doers and often lived to a very old

[11] See Chapter 6 on Spanish horses.

[12] See Chapter 8 on the Thoroughbred horse.

age. They also had plain heads and strong necks, long fine bodies with strong backs.

Still in Morocco, where there were reckoned to be at least 400,000 horses, these horses were "well bred and generally of Arab origin" but they were more elegant from head to shoulder and breast than the Arab itself. The horses of Tingitana, says Ammon, were greatly valued for their speed and courage and although the Moors are not as careful breeders as the Arabs (they are, of course, not themselves descended from such an ancient race of horsemen), their horses were amongst the best in the world for endurance in heat, cold, hunger and thirst.

The reason why some Berber horses acquired such disrepute in Western European eyes lies perhaps in the unhorsemanlike way in which they were broken in. The method used to break twelve to fifteen-month-old yearlings frequently gave them straight shoulders and hard mouths.

Some writers of two centuries ago mention that by far the best horses were to be found in the province of Sus, in the south of Morocco. "I found that those good qualities of the Berber horse of Arab descent which I had thought exaggerated to be well below the truth because they are really the most courageous and easily managed horses to be found anywhere." This writer then describes a really perfect horse, although it possesses a markedly convex face! Many were of a beautiful cream colour with black hairs intermixed. "Nothing could equal the fire, agility and docility of these remarkable horses," says a traveller called Riley, and he describes an event in which 1,200 Moors and Arabs took part, all mounted on stallions. He further adds that they would have made some of Europe's best racehorses look like cart-horses (Ammon).

Even so, the quality of horses in Morocco declined, although they still retained their incredible hardiness. The ordinary horses of Tunisia and Algeria were very similar to the ordinary Berber. Little attention was paid to breeding and very little care given to them, although a strain of more quality was to be found in the Tunisian interior in the Biledulgerid province. These horses were bred by the wandering Bedouin and were said to be very good-looking. In Tripolis, breeding was done with great care. Horses had the speed, strength and excellent conformation characteristic of a really good horse, yet they were only 14 h.h.

In Dahomey, Equatorial West Africa, there was said to be a considerable number of horses which belonged to two different breeds. There was the descendant of the Nigerian horse which made up about 20 per cent of the equine population, and a native horse that was called after the Berber people.

The Berber horse was, and so far as I know still is, small and brown in colour, and stands about 12·3 h.h. (1·30 Stm). About 50 per cent of this breed are of a light bay colour and even smaller, standing only about 11 h.h. This strain of ponies is strong and hardy. They live almost in the wild state, although the natives use them for riding and also for draught purposes. They are extremely resistant to sleeping sickness which the Nigerian horse is not. Crosses between the two breeds result in an intermediate strain which possesses the good points of the Berber pony.

The Beni-Moussa tribe lives in the Middle Atlas Mountains and the Ait-Attab to the south-east. The temperature of the climate is unusually high with strong variations and low rainfall. Both tribes are famous for their knowledge of horse breeding. The horses are the result of cross-breeding. The predominating type had a convex face and sloping croup, another a straight face and flat croup—the Berber plus Arab type—whilst a third possessed a pronounced ram's nose and rather weak quarters. The result of this parentage is a horse with a fairly heavy head, convex profile with very pronounced eyes, an arched neck, rounded back, pronounced withers and flat croup. Also a high, rather flat, breast. They often stand badly!

We are now able to appreciate the presence from prehistoric times of the ancestor of the Berber horse in North Africa. The Mauretanian tradition extended to both Africa and Europe (Spain) as early as 10,000 B.C. and there was an active bond between the inhabitants in the fourth millennium B.C. Although there may well have been a coastal route along the North African coast to Egypt and agriculture was practised from Mesopotamia, Syria, through Egypt to Tunisia, there seems to be little evidence of an interchange of peoples.

It is generally held that the peoples of the early dynasties in Egypt did not know of or use the horse. They certainly did use asses and they also had trading relations with Akkad, Aššur and Elam. I am, therefore, not so certain that a few centuries later, when the domesticated horses appeared in Mesopotamia and Anatolia, which we have already discussed, this animal remained quite unknown to the Egyptians prior to the termination of the Hyksos rule. For it is the arrival of the Hyksos, the word is Egyptian for 'foreign invaders', in 1670 B.C., which is the general but erroneously accepted date for the arrival of the horse in Egypt.

The expansionist Churrians or Hurrians were largely responsible for the movement southwards of much-mixed ethnic groups of Near-Eastern peoples, to use the modern term, 'displaced' from Syria and Palestine by the Hurrian takeover. These displaced persons had formed themselves into an organized army and finally settled down in the delta of the eastern Nile to rule Egypt. Although the Hyksos were Mesopotamian–Asiatic in origin and must have had some knowledge of horses, there is apparently no evidence that they actually did bring horses with them. Such evidence is often based primarily on burial finds.

The Hyksos rule began with the Fifteenth Dynasty but it was first, with the discovery of rich tombs of the Eighteenth Dynasty, that the famous Egyptologist W. M. Flinders Petrie found the remains of horses and asses together with those of the dead. This, however, is at the end of the Hyksos power when they had begun to introduce improvements through their connections with Asia. They may even have asked for war chariots to defend their failing power. And it is at this period that the first horse-drawn chariots were imported and with them horses in any number.

As a result of the Hyksos occupation, Egypt not only changed her habits with regard to warfare and other technical advances but also her internal political organization and mental character. From now on there began what we may perhaps call the age of chivalry in Egypt.

As early as the nineteenth century B.C. Ras Shamra and Ugarit, Syria, were probably politically dependant upon Egypt, and in the Thirteenth Dynasty Intu, King of Byblos, paid homage to Neferhotpe of Egypt. After the Battle of Nig during the sixteenth century B.C. Thotmes III of Egypt, who appears to have driven a chariot made of gold, received as tribute 180 horses, but after conquering Megiddo at least 2,041 horses and 191 foals or yearlings were taken. He also received animal tributes from many smaller countries once he had secured the Phoenician ports. His grandson Thotmes IV married a Mitanni princess and in this way he was able if he wished to obtain a constant supply of horses from Mitanni. Tell Chagar Bazar, which has already been mentioned, lay within the kingdom of Mitanni and for hundreds of years had been the largest horse market in Asia.

I have come across no references at all to specific studs in Egypt. Perhaps there were none on the organized scale of Mesopotamia and Asia because the Egyptians could so easily obtain and import all the horses they needed. From all the evidence before us, it seems plain that the same strain and type of horses had spread throughout Syria, Mesopotamia and Egypt. There was considerable trade between these countries apart from the friendly relationships which existed between rulers, as we have seen from letters.

Naturally, too, there were horses captured in war, which was both a hobby and a way of life for many rulers of ancient kingdoms and it was one which they were able to indulge in often on a massive scale, once they were sufficiently advanced to have trained war horses.

If we really want to know what the conformation of this general strain of eastern horses was like, we should remove from the Assyrian, Babylonian and Egyptian representations everything that has to do with the *style* of the representation. We then have a very general picture of the small, elegant oriental horse. If one looks at these representations through the eyes of the artists, then basically they all show us the same small, well-bred animal.

In discussions of breed history, one must not compare the Egyptian horse with the Persian,[13] only the representations of horses of similar

[13] Although the Nisaean and Ferghana horses and the Persian cob were individual breeds and are discussed later.

origin executed by different artists! For example:

Syrian – Egyptian horse in Egyptian representation.

Syrian – Assyrian horse in Assyrian representation.

Syrian – Iranian horse in Persian representation.

Syrian chariot-horse Persian-style. Xerxes palace in Persepolis, *c.* 480 B.C. This horse has a straight face with a curve over the eyes and may have been of Cilician origin.

Iranian saddle-horse Persian-style. This horse is larger than the chariot horses and has a convex profile. Apart from this animal we have the generally distributed oriental strain. One must observe carefully the costume of the people bringing tributes. If in Syrian dress, then the horses are probably Syrian—or from Asia Minor. If the dress is Persian, then the horses could be from Media, Iran or Bactria. Xerxes was King of Persia (485–465 B.C.)

In the seventh century B.C., a few centuries before Herodotus (of whom we shall hear more) was born, there appeared on the northern Asiatic steppes a nomadic people comprising many tribes called Scythae in Latin and *skuthai* in Greek, which simply means 'nomad' and cannot be accepted to mean either a people or a breed of horses. The Persians called all Scythians *Saka*;[14] their language was, in fact, Iranian, and their original home to the north of Bactria and to the east of the Amu-Darja delta on what is today known as the Kazakstan Steppes.

They spread far and wide—to south Russia (Transylvania and Bulgaria) and into Asia—on their fast horses.

Both the Massagatae who lived to the north of Chorasem, west of the Aral Sea, and the Sacae were Scythian peoples. "The enemy, mighty with his horses and his swiftly flying arrows, lays waste the countryside far and wide." (Ovid, first century A.D., writing from the shores of the Black Sea: *Trista ex Ponto*.)

The Don Scythians appeared on the South European Russian scene. They were the ancestors probably of the famed Don Cossacks. Their power was the saddle, and briefly what happened was that the restless Scythians upset the Cimmerians, who in turn upset the Thracians, and thus the pressure of mounted nomads moved slowly westwards. It was this pressure that destroyed both early Europe and the ancient Orient. It was due entirely to the mobility which the horse provided and this, in turn, relied upon a constant supply of horses from breeding centres. "A band of Scythian farmers quarrelled with their fellows and withdrew into the land of the Medes," says Herodotus. But they also drove waggons, carried their houses about with them (hooded waggons = *araba*), rode on horses and shot with the bow.

Whilst the Assyrians ruled over upper Asia, the Medes unsuccessfully fought the invading Scythians who spread themselves over all Asia. They

[14] *Saki* means 'deer'—the people of the deer.

even went against Egypt but were turned back by gifts. Here one may perhaps remark that the Scythian way of life and their knowledge of horses was *very* like that of the latter-day Arabs. Somewhere, however, they did not have it all their own way because (as Herodotus says):

> Scythia produced neither ass nor mule, because of the cold and the asses confounded the horses of the Scythians because of their braying.
>
> The Scythians would leave a part of their own herds behind with the keepers and ride away themselves to another place. Then the Persians would take the herds and would be pleased by what they had done but Darius was in a great strait at last, having acquired so many herds. And the Scythians sent him a bird, a mouse, a frog and five arrows. Darius thought they were giving him earth, water and their own submission. Mouse lives in the earth eating corn as men do, frog lives in water, and a bird is like a horse in swiftness. The arrows were the strength of the Scythians who were surrendering.

But a man called Gobryes interpreted it otherwise: "O, Persians, if ye become not birds and fly up into the heavens, or mice and descend beneath the earth, or frogs and leap into the lakes, ye shall not return home again being smitten by these shafts."

> Now the Massagatae wear raiment like the Scythians and have the same way of life. They fight both on horseback and on foot and with bows and with spears and are wont to carry battle axes. Gold and brass they use for all things; for their spears and arrowheads and axes they use brass, but their heads and girdles and breast-bands they adorn with gold; and likewise about the chests of their horses they put breast-plates of brass, but adorn with gold the bridles and bits and bosses. Iron and silver they have not in their country but gold and brass abound.
>
> For that which the Greeks say the Scythians do, it is not the Scythians but the Massagatae. They are drinkers of milk [live on their herds—nomads]. The only god they worship is the sun to whom they sacrifice horses; and the meaning of this sacrifice is this, that they apportion to the swiftest of the gods the swiftest of all mortal things. (Herodotus)

By about 500 B.C., the Scythians had penetrated north of the Borysthenes and had become settled farmers. To their south were more or less settled Scythian herdsmen, whilst the 'royal' Scythians occupied an area on the west bank of the River Tanais. To the south of them around the Black Sea and to the east over the steppes the nomad Scythian tribes wandered. All came under the denomination Indo-European, which included the west Celts, Germanen and Celtiberians.

Interestingly enough, at Carcintis, a small Greek occupied territory on the north shore of the Black Sea above the Crimea, there was a much-frequented race-course.

The Scythians called themselves Skoloten.

> When Xerxes the Persian went against Greece there had to be organized ships to carry horses. The king was preceded by one thousand horsemen

chosen from all the Persians, and then by one thousand spear bearers, chosen from them all with the points of their spears turned towards the ground, and then by ten sacred Nesean horses, adorned exceedingly richly. Now they are called Nesean for this reason: there is a great plain in Medea named Nesaean, and this plain produces these great horses.

And behind these ten horses came a chariot sacred to Zeus, drawn by eight white horses; and behind the horses the driver followed on foot, holding the reins; for no man mounts upon that seat. And behind this chariot was Xerxes himself, in a chariot drawn by Nesaean horses; and the driver whose name was Patiramphes, the son of Otanes a Persian, stood beside him. So Xerxes rode out of Sardis but whenever he was so minded he got down out of the chariot and rode in a carriage. Followed by another thousand horsemen. Then ten thousand spearmen and another ten thousand horsemen, followed by the host. (Herodotus)

We have considered the type and strain of horses that were domesticated in the Near and Middle East and China but we have not yet dealt with the strains which evolved in ancient Bactria—now part of Afghanistan and in Sogdia to the north. Bactria, at least, lay within the kingdom of the Medes and Persians *c.* 800 B.C.

As we have seen, larger horses were bred in Urartu and the forebears of these horses could originally have come from Europe via the Pontic–Caspian steppes—the *kobyla* of eastern Europe. I think this is the only way that the heavier type could have been introduced to Asia. When crossed with the local oriental strain, they would produce a short-barrelled, short-legged horse.

The famous Nisaean horse was raised on the Nisaean Plain—at no great distance from Urartu. This plain is situated in Iran, formerly part of Media. Modern Teheran appears to be situated upon it. From illustrations it is clear that the Persians rode two types, the cob and a larger middle-weight horse. The Nisaean was the latter type.

Not only selective breeding but fodder as well makes for size. Lucerne was grown in these areas and was popular fodder for horses. As it contains chalk and will grow only on chalky soils, this is also a constituent to help raise bigger-boned animals.

To the north, however, there was yet another Nisaia, geographically mentioned around 185 B.C. and lying in Parthia which was part of Media.[15] And large horses were, at this period, bred in this area too. There were breeding districts or studs at Merv and Bactria (Balch),

[15] The Assyrians did not distinguish between Medes, Persians or Parthians. Like the Greeks, the Assyrians called all of these peoples *Madai*, Medes. Neighbours of the Medes were Assur, Urartu (Armenia) Mannae and Ellipi (Iraq?). The latter had an autochthonous population although one of the king's sons had the Iranian name Ispabāra, 'horseman'. Colt and stallion are incidentally two of the animal incarnations of the god Varhran, who is the Iranian Indra. The Parthians were a Saka tribe, forerunners of the Saka migration of 130 B.C., subsequently becoming Iranian.

39. Athenian mounted warrior on a sixth-century B.C. monument

40. Greek horseman, *c.* 550 B.C. From Grumentum, Lucania

whilst in Ferghana larger horses which were just as famous as the Nisaean were also bred and were 'exported' to China. The modern Turkmene breed is the Achal Teké of the Ashkabad stud in the same area.

There also seems to have been a smaller pony type bred around Hyrkania on the east coast of the Caspian Sea, and if a charming gold model of a quadriga team of four is anything to go by, there were also smaller horses which belonged to the Chorasem area. It seems probable that the strain was known as Tochar and that it has continued to survive in the recently discovered Caspian pony of the Elburz Mountains.

Herodotus[16] wrote of India: "the living creatures are far larger than elsewhere, excepting the horses which are surpassed by the Nesaean horses of Media", Of King Darius, the Persian king, he says that this king could put an army of 700,000 in the field and many were horsed. In fact, Darius's cavalry were 'internationally' famous. It is interesting that Herodotus further mentions tribute paid to Darius who received from the Cilicians "three score and three hundred white horses, one for each day of the year, and 500 talents of silver".

This area in Media continued as a horse-breeding locality right into Sassanidian times (A.D. 226–637). Ardashir I, a Persian ruler and a descendant of Săsăn, founded his kingdom by conquering Parthia and Media. Antonius tells us that the stud was situated near Harunabad, south-west from Kermanshah. But he questions if Iranian horses were generally larger than the horses from Syria or Asia Minor. In his *Anabasis* Xenophon says that his captured Armenian horses were smaller than the Persian but a lot faster. This statement rather bears out my theory of the breeding of the Persian horse, whether the cob type or strain or the larger animal. From five to twenty years of age Persian children were taught to ride horses, to shoot with the bow, and to tell the truth.

Herodotus also gives us snippets of information about other peoples and their horses, and since they bear on our theme, I give them here. He says that the Greeks learned from the Libyans how to yoke four horses together. In Athens it was a sign of great wealth to keep a four-horse chariot. Militrades, an eminent Athenian, was of a house that kept four-horse chariots and he had won a victory at Olympia with his horses. One can imagine the tremendous excitement nearly 2,400 years ago in the household of Militrades, the same kind of excitement as would be found in any British, American or Continental home when the scion brings home an Olympic gold!

Artybuis, too, rode a horse that was taught to rise upright against a

[16] A Greek historian of 484–425 B.C. He was born at Halicarnassus, a Persian dependency. As a young man, Herodotus travelled extensively. Although his writings are sometimes questioned, I feel that his information about horses is probably generally true.

foot soldier, and young Athenians generally were keen to perfect their own schooling as well as breed and raise good horses.

Greece, of course, is not a country which lends itself to raising horses. The country is too mountainous and the plains are too small. But of the horses which were bred, the Thessalian was renowned for beauty, courage and endurance. Possibly the forebears of this breed were imported from Cilicia or some other place in Asia Minor, and were crossed with the European 'cob'. Bucephalas (Bull-head), the charger of Alexander the Great, was one of the best of the Thessalian breed. He was black with a white star, and very large. He had enormous stamina as we can gather from the manner of his death at the age of thirty. He is the only horse to have had a city—Bucephalia—founded in his honour.

Other breeds of horses that were favoured by the early Greeks were the Argive, Acarnanian, Arcadian and Epidaurian. Unhappily there is little information about their finer points of conformation. However, the poet Oppian, who lived a few centuries later than the period presently under review, was a great believer in the power of the horse's eye, holding that the eye of the steed should match the eye of the quarry in colour—bluish for bears, yellowish for leopards, fiery and flaming for wild boar, and flaming grey for lions.

Now as to colours, the Thessalian was usually white or grey, the Thracian was white, the Argive was bay or chestnut. The Tuscan, Archaen, Armenian and Cappadocian were recommended for hunting and as chargers, whilst the Nesaean[17] was "of exceeding beauty and docility", having a full, lightish-coloured mane.

Young Athenian horsemen probably based their ideal type on the cavalry horse shown so beautifully on the Parthenon frieze which is well described[18] as a "cobby but spirited and corky horse, with a light and somewhat peculiar carriage of the head and neck, just as we see represented on the Elgin marbles".

Simon (*c.* fourth century B.C.) mentions three sizes: large, small and good-sized, or moderate, which is the best. He also says that the horses from Thessaly are the best and goes on to describe the points of a good horse with which all true judges of horses will agree.[19]

Referring to other peoples, Herodotus tells us that the Persians even fought beyond the Ister (Istros), as well as going as far as Libya.[20] He says:

[17] Armenian Cappadocian and Nesaean—a further indication that horses were imported into Greece.

[18] "Stonehenge" wrote on the horse, according to Xenophon's *The Art of Horsemanship*, trans. by M. H. Morgan, Ph.D., in which book will be found descriptions of equine conformation by Varro, Vergil, Calpurnius Siculus, etc.

[19] See Xenophon.

[20] The prisoners of Barca, Libya, were given a village in Bactria which they also named Barca.

41. Relief from a Greek tomb from Xanthos in Lycia, Asia Minor, showing a funeral procession—*c.* 470 B.C. Large and small strains of horses are depicted

42. A carving of horsemen with a cob, from the Parthenon, Athens, *c.* 440 B.C.

43. *(bottom right)* A marble relief of the goddess Hecate with a pony, from Thessaly—*c.* 350 B.C.

44. *(below)* Horse of Selene, from the Parthenon, Athens—*c.* 435 B.C.

> Beyond the Ister there are the *Sigunae*. They wear Median clothes. Their horses have hair all over their bodies to a depth of five fingers breadth, are small and have flat noses and are able to carry men but are exceeding swift under yoke.[21] So that the people of that country always rode in chariots. And their borders extend nearly to the *Eneti* on the Adriatic—they say they are colonists of the Medes.

On page 116 I mentioned Chorasem on the south bank of the Aral Sea and the small strain of horses which were bred there on the delta of the River Amu-Darja (Oxus, illustration 12). The people of Chorasem had inherited the Kel'te Minar tradition of the third millennium B.C. This was a settled hunting community which eventually, but not at first, bred sheep, domesticated cattle, and had the onager to hunt as game together with wild horses. It appears that it was the Siberian Afanasjevo civilization that first bred horses.

The next stage was the Taza-Bag-Jab civilization and here Tolstov notes that the latter combined with the Iranian tradition of the origin of the mythical first man from the north (note especially the north!), his crossing of the sea, the delta of the Amu-Darja and of Chorasem as the place where he first lit the holy fire. On the other hand, the Su-Jarghan combined with the Chorasem legend handed down by al-Biruni, in which it is said that during the thirteenth century B.C. Siljavuš, the mythical ancestor of the Chorasem ruling dynasty, wearing a golden helm and riding a black horse, came from the south (Iran?) to the Amu-Darja and founded the kingdom of Chorasem. This area, including the northern district of the Aral Sea, was also known as Turan (modern Turkestan) and we sometimes find references to the Turan horse. I cannot locate it as a breed, but it is undoubtedly the ancestor of the Turkmene.

The knowledge of horse breeding probably spread then from central Asia through the Andronovo civilization, which was responsible for Asia's Bronze Age prosperity in 1250–750 B.C. This civilization had connections north to the Sejma and west to the Pontic Tripolje civilization referred to on pages 80–81. The period mentioned above was a dynamic time for improving methods of cattle breeding and farming. The later Andronovo peoples were also the precursors of the early nomads (see Chapter 2), chronologically pre-Sacae, continuing unconditionally until *c.* 700 B.C. and including dimensionally a number of later famous horse-breeding centres within its own areas of expansion.

Nomad pasturing, which was still practised by the Kazak nomads until they relinquished their nomadic way of life on joining the U.S.S.R., is only possible with the help of horses. In high winter when the ground is covered with snow and frost, horses scrape the snow away from underlying grasses, cattle follow, and after them come sheep and goats. Horses

[21] Unless this is merely an expression—to yoke a horse, meaning to put one to—it is surprising that the yoke is still used. The choke collar had long since replaced it in Asia.

also provide the necessary means of transport for long-distance change of grazing, and they are also used for racing and hunting and flying the eagle. "The nomad certainly could not live without his horse and when a man died his horse was buried with him—and the horse was honourably killed with a military weapon, the battle axe."

In a mountain valley in Ferghana (east Turan) there are 91,900 petroglyphs of wild animals cut into the sandstone cliffs: maral deer, mountain goats and sheep and hunting scenes of people with domestic animals such as dogs, cattle and yak. The yak, of course, is still used under the saddle and as pack transport. But alas! there appears to be no indication as to the part that horses played in this, one of the historically most important areas of horse breeding at this period. In the south-east Ural steppes, 1700–1200 B.C. horses were apparently only used for meat and were a small, strong strain. In the Kurgan burials, 1200–700 B.C., horses held a markedly high position and were of higher build with finer legs than the Ural Fedorovsk medium pony type.

A fairly accurate reconstruction of the sort of horses for which ancient Turan was famous has been made from the frost-conserved remains found in the Pazyryk tomb burials. A number of European and Russian archaeologists such as Gryaznov, Rudenko, Jettmar, Vitt and Hančar have described the horses found in these extensive tombs in which chieftains and aristocrats were buried. According to Gryaznov most of the horses found in the kurgans "belonged to the local steppe breed, of Mongolian type—small shaggy heavy-boned beasts, typical of the steppe horses that lived in natural herds on open grazing all the year round". But the horses buried with the tribal chiefs were a quite different breed. These were "tall animals with a lean and rather aquiline head set on a long neck. They were thoroughbreds of the same type as the chargers of Central Asia which were so widely famed in ancient times, and were fed on corn and tended with particular care".

Referring to the peoples of the Altai during the fourth to fifth centuries B.C., Gryaznov adds:

> The dominant factor was that all the warriors of the tribe were mounted. They are represented on gold plaques from Siberia with *gorytus* hanging at the waist containing a short bow (100–110 cm) like that used by the Scythians, and short arrows (50–60 cm). With a bow of this kind a horseman was able to shoot while riding at full tilt, directing his arrows to the front, to the rear or to either side as required.

Herodotus states that the Scythians fought on horseback with bows and arrows and on foot with spears and daggers. The military tactics of the nomads seem to have been to charge headlong against the enemy, firing from their horses as they galloped.

To return to the Pazyryk tombs. Professor Vitt[22] examined the horses

[22] The late Professor V. O. Vitt was a leading hippologist, holding the chair of hippology at Moscow University.

found in the Pazyryk tombs and also those in the Sibe kurgan in the Altai Mountains and placed them in four groups.

Group I—145 Stm., 14·1 h.h.
Group II—140 Stm., 13·3 h.h.
Group III—136 Stm., 13·1 h.h.
Group IV—132 Stm., under 13 h.h.

Groups I and II were typically fine-boned and belonged to a saddle-horse strain: "having a light head, concave profile, short jaw with lengthened diastema, a long neck, short croup and forelegs longer than hind legs" (Hančar). The variation between the best type in Group I and a specimen in Group IV was considerable and it may be accepted that they belonged to different breeds. The horses of Group I were of a selectively bred strain, probably imported from Turan and descendants of the 'heavenly' horses of Bactria (see index). This type has been likened to the modern Achal Teké. And it is of particular interest, from measurements of the skeleton, that Professor Vitt assessed a horse belonging to Group III (13·1 h.h.) as resembling the well-known saddle-horse with leather stirrup strap of the frieze on the Certomlyk Kurgan silver vase from the fourth century B.C. This frieze proves that the idea of a stirrup, in this case a looped thong, is much older than is commonly supposed.

Furthermore, Vitt was able to show that the tremendous improvement in growth of the Pazyryk breeds was due to castration of fairly young horses from eighteen months to two years. For it is suggested that castration increases the period of cannon-bone growth and helps towards the development of finer bone; it decreases ossification, and although it makes little impression on development of either quarters or spine, it does improve the length of neck and limbs—especially of the forehand, less of the hindquarters. Thus the desired 'quadrat' shape of the ideal saddle-horse is achieved. By this means, says Hančar, in the Altai Mountains too, a high standard of breeding was possible and became an historic reality during the first millennium. That breeding was widely distributed is seen by the fact that the ear-marks of ownership varied with each of the horses found in Kurgan 1—and this is where the idea of the ten 'studs' arose. It is more likely that these were the marks of tribal ownership.

Generally the height of the ordinary steppe horse, then as now, was about 13·2 h.h. and the subsequent burials in the fifth to fourth centuries B.C. and third to first centuries B.C. show an improvement to a type of fine-boned horses from 14 h.h. to over 14·2 h.h.

The best Pazyryk horses belonging to Group 1 were male and stood between 14·2 h.h. and 15 h.h. with a long neck, fine head carried high, small browed and marked concave profile, a short back, fine strong legs, well-developed joints, which certainly resembled the present-day Arab type, although 'significantly commoner, heavier with more bone and a thicker coat', generally a dazzling bay/chestnut colour ranging from liver or dark bay to bright golden bay with black mane and tail.

None of the horses had any white marks or feet which Vitt stresses as an important point showing the breeders' experience of the value of selection since white hooves are always more brittle and without shoes are useless on a riding horse.

We have already noted the preference of some peoples for horses of distinctive coat colours. Vitt notes that at the time of the birth of Christ, both Greeks and Romans preferred brown and grey horses. The Karst horse of early times was probably predominantly grey and the colour preference of the Huns *c.* 200 B.C. is shown by Shanyü Maotu with symbolic magical design, in encircling the Chinese army near Peteng, using grey horses for his eastern cavalry, bay for the southern army, white for the western, and coal-black horses in the north.

But as Hančar remarks with further reference to the golden bay Pazyryk horses, there is an apparent bridge between the Altai horse-breeders of the first millennium B.C. and their colleagues in Central Asia whose preference for bright bay horses is connected with sun worship. This cult was practised not only by the Bactrians and Massagetae but also by the Medes, Persians and Indians and also much further north, as we see from the Trundholm chariot (illustration 31).

In Pazyryk Kurgan 5 there were, beside four saddle horses of the type mentioned in Groups I and II, a harness horse of Group III, and a parade carriage with a team of three horses similar to the Oxus horses of Group IV. All four harness horses were small and showed less quality but they were so well conserved by freezing that it was obvious that not a single horse had been overworked or showed any other sign of wear and tear.

Gryaznov is of the opinion that the large felt carpet found in Kurgan 5, with the horseman paying homage to the figure of a goddess, was imported from somewhere outside the Altai area. Whether it came from the

> nomads of the steppes, around the Aral Sea or the deserts of Sinkiang, from Mongolia or Ordos—cannot be solved until the decorative art of all the nomadic tribes has been thoroughly studied as that of the Altay peoples. In the meantime the existence of cultural interchange over a wide area is demonstrated by the rapidity with which numerous characteristic features of the nomadic culture spread throughout the whole steppe region. (Gryaznov)

Between the Danube and the Great Wall of China similar forms of weapons, harness and animal styles are found. If we analyse these findings from the written material available it seems that these kurgans produced at least four breeds of horses: Group IV the small Tochar; Group III the ordinary mountain pony; and Groups II and I saddle horses ranging from the type commonly used on the steppes by the Scythians to a better bred sort similar to the present-day Achal Teké which may have been the selected type reserved for the nobility.

From the foregoing text we have seen the emergence of the Oriental or Eastern horse. There is no doubt but that the Inner Asiatic steppes provided the earliest ambition to breed domesticated animals. From the earliest form of selective breeding within the herd system we have seen how special areas and studs for horses evolved and developed throughout the Middle East and China and how some of these were devoted to specialized selection for colour or for ability. Without doubt it was from these strains of Oriental horses that the nomad Bedouin acquired their horses.

Now we come to a strain of horses which has been the most debated and argued over in all the history of horse breeding, namely the Arabian. Antonius writes: "From ancient writings we know that the Arab himself took possession of such horses quite late, certainly not much earlier than Anno Domini; so the description 'Arabian' horse is so far only substantiated by the fact that it was the Bedouin who kept the old strain which they had taken over."

Geographically, Arabia or Aribi (a century or so earlier it was called Aramu) appears to the south of Syria in the fifth century B.C. and *within* the kingdom of New Babylonia, and the same area is called Arabia Deserta in the fourth century B.C. Assyrian inscriptions mention the Arabs. A tombstone from Hauran dated A.D. 328 is said to bear the name of one who called himself "king of all the Arabs". But in fact Arab tradition goes back only a few generations before Mohammed, who died in A.D. 632. Of Mohammed "Nothing is known with certainty about his early life, though imagination has been busy." This rather sums up the position of those horses which were imported into Arabia and which have been defended with so little reason. If conformation and stamina do not of themselves recommend a breed, then partisan prejudice can only cause doubt. To these early strains a number of names were supposedly given by Mohammed. Nothing can be proved or disproved as to where the horses originated—they could as easily have come from Egypt as from Syria. But they were most certainly not native in the accepted sense.

The Bedouin—the Ahl-el-Beit, the people of the tent—were free nomadic peoples of many different tribes. Probably some were the descendants of earlier Asiatic nomads. Until these Bedouin peoples had domesticated the camel, which happened about the sixth century A.D., they could not make much use of the horse. So it follows that the earliest Oriental horses which were brought into Arabia by nomad Bedouin could as easily have been called Bedouin as Arab.

It is so often asserted that Arabia is and was the native land of the Arab horse. Unfortunately there is no real evidence for either a native population of men or horses prior to Mohammed. The subsequent numerous strains of horses obtained their family denominations apparently from the fact that they were owned and bred by the Moslem population known to us as Arab and it is a case of a breed of horses of many and varied families and types that were named after a race of people and not a country.

The earliest actual information about the horses of the Arabs is provided by the Greek poet Oppian in the second century B.C., who recommends them for hunting, and it would appear, too, that there were not many horses in this land since Arrian tells us that horses were being imported to the south from Egypt and later these were known as Sapphar horses from the name of the present district of Dafar.

However, the Saracens from the north, in the fourth century B.C., *did* ride horses. Arrian describes them as light and plain but with plenty of stamina, although generally the Arabs kept to their camels. Now some of the Bedouin appear to owe their ancestry to the Saracens who were true nomads, knowing nothing of agriculture and occupying themselves only with war and hunting.

The Arabs, however, had horses a century or so prior to Mohammed and it seems fairly obvious, if comparisons are made, that the likely source would be the descendants of the ancient Syrian and Assyrian chariot horses, whose ancestors, as we have seen, originated in the steppes of Inner Asia. Coming themselves also from an ancient equestrian stock with an inborn love of horses, the Arabs lost no time in improving the strains. Indeed, if one depends for one's life on one's horse, then only the best is good enough. Therefore to breed the best was a necessity as well as an art.

The Bedouin, or Bedavi, who bred and possessed the most beautiful horses come into the picture at a time much later in history. And although the Arabs themselves trace their legendary creation to Ismael, the son of Abraham who lived about 2000 B.C. and believe that there were *then* horses in Arabia and that the noblest horses are descended from a stallion from King Solomon's stables, they have made an error in time. As we know, Solomon lived in the tenth century B.C., and incidentally, Solomon bought his horses from Egypt and Cilicia.

There is a charming story attributed to a work by two Arabs, Abu Abdallah Muhamed Bin Sajad and Abu el Mundir Häshan Bin Muhamed Elhälebi Elbalassi, regarding the present of one horse to the Queen of Sheba and from this horse there sprang a number of strains, none of which could be identified under any of the then existing strains 150 years ago.

Solomon himself was the third king of Israel and reigned in the tenth century B.C. All the important caravan routes to the north and south went through his kingdom. In Megiddo archaeologists discovered a complete complex of boxes and stables which formed part of the royal mews. There were boxes for at least 450 horses and coach houses for 150 chariots. In I Kings 10:26–29 we read:

> And Solomon gathered together chariots and horsemen: and he had a thousand four hundred chariots, and twelve thousand horsemen, whom he bestowed in the cities for chariots . . . and Solomon had horses brought out of Egypt . . . and a chariot came up and went out of Egypt for 600 shekels of silver: and an horse for 150: and so for all the kings of the Hittites, and for the kings of Syria, did they bring them out by their means.

But horses were also obtained from Koa[23] which apparently was a part of Cilicia. Cilicia was equivalent to the Newmarket of modern times. There may be an exaggeration since the power of the Hittites was waning by the tenth century, although the text does infer that good horses had still to be imported into Anatolia, the climate being unsuitable for breeding good horses.

At this point the Queen of Sheba (Saba), who ruled at Marib in the present country of Yemen, began to trade with Israel. At first the only method of transport was by pack-donkey slowly from spring or oasis to spring, to water the animals—all 2,000 kilometres (1,200 miles) from Marib to Ezeon-Geber. But then the camel was domesticated and goods could travel much faster. There is no mention of horses in Arabia Felix, the countries of southern Arabia. Although when the Queen of Sheba herself visited Solomon to discuss matters of trade, he gave her all she asked for: "beside *that* which Solomon gave her of his royal bounty. So she turned and went to her own country, she and her servants". In the Bible there is no mention of horses, but one can scarcely credit that someone so active as to undertake a 4,000-kilometre journey by camel would not have seen the advantage of asking for horses.

Historically Yemen is one of the few countries in Arabia where horses were eventually to be found and these ironically enough were Cappadocian imports.

It is interesting to note that the Persians called the Syrians Cappadocians, and so it makes sense when we learn that in A.D. 344, three years before his death, Constantine the Great presented the king of the Homerites who ruled over part of Yemen (once the kingdom of Saba!) with 200 horses from Cappadocia "which were then just as famous as the present-day Arab". (Ammon) Thus even in southern Arabia where we might expect to find a native primitive horse, if this were geographically possible, evidentally there were no horses. Although today, Yemen has three definite strains of Arabian horses.

Of course, there may have been some horses in Arabia at an earlier date than can be established but they were certainly few. Even if the existence of the prophet Job is doubted, it is remarkable that amongst the inventory of his possessions, sheep, oxen, asses and camels, there are no

[23] Werner Keller, *Und die Bibel hat doch Recht.*

horses. And in David's time when war was made between the Israelites and the Arabs and the former won, their booty consisted of slaves, camels, sheep and asses—but no horses! The Bible contains many similar references but I have not found one to the fact that the Arabs had horses. Even Herodotus (*c.* 450 B.C.)[24] tells us that when Xerxes, King of Persia, went to do battle against Greece, Arab riders accompanied him and also fought—but they all rode camels.

Quite a number of Asiatic folk had to pay tribute to the kings of Persia and those who bred horses paid up with horses—but Arabia's tribute was incense. It does not seem likely that Persia would have allowed Arabia to get away with 1,000 pounds of incense if such beautiful horses were bred there *at this period* as was the case at a later date.

Strabo wrote of Arabia Felix: "there are no horses, asses or pigs", and of the deserts: "the country has no horses, there are camels in place of them".

The language of the Arabs themselves is Semitic and the Semitic languages are grouped, thus showing the probable origin of these people and therefore their connection with horses and their ability subsequently to breed and improve on the strain.

1. North-east Semitic or Accadian (Assyrian and Babylonian) which is said now to be extinct.

2. West Semitic or North-west Semitic comprising Early Canaanite (possibly Ugaritic). Hebrew and the Syriac tongue.

3. South Semitic which is apparently subdivided into north Arabic or Arabic which is spoken by about sixty million people.

Not being a student of peoples as opposed to horses, I am not in a position to judge which of the Semitic languages was used by the Bedouin when they first came into contact with horses. I have given examples of the Chinese, Cassite and Hittite words relating to the horse before the birth of Christ. But there appears to be no historic reference in the Arabic language to the horse—or if there is, I have not met with it—in translations in any of the European languages with which I am familiar. It is for this reason that I view with suspicion the pretensions of breeders who claim that the desert Arabian carried his pedigree around his neck. How many followers of Mohammed could read let alone write? The existence of Mohammed's cherished mares is more legendary than factual. I have seen at least three different versions, the mares having varying names and belonging to varying families.

Throughout the centuries the Bedouin improved the strain or strains, for there were many. Undoubtedly he had a traditional love for or understanding of horses, but it was necessity that led to perfection through selection. He could not afford to own a bad horse. Whilst the average Bedouin in any decade owned only one horse, sometimes several men

[24] See Herodotus's journeys.

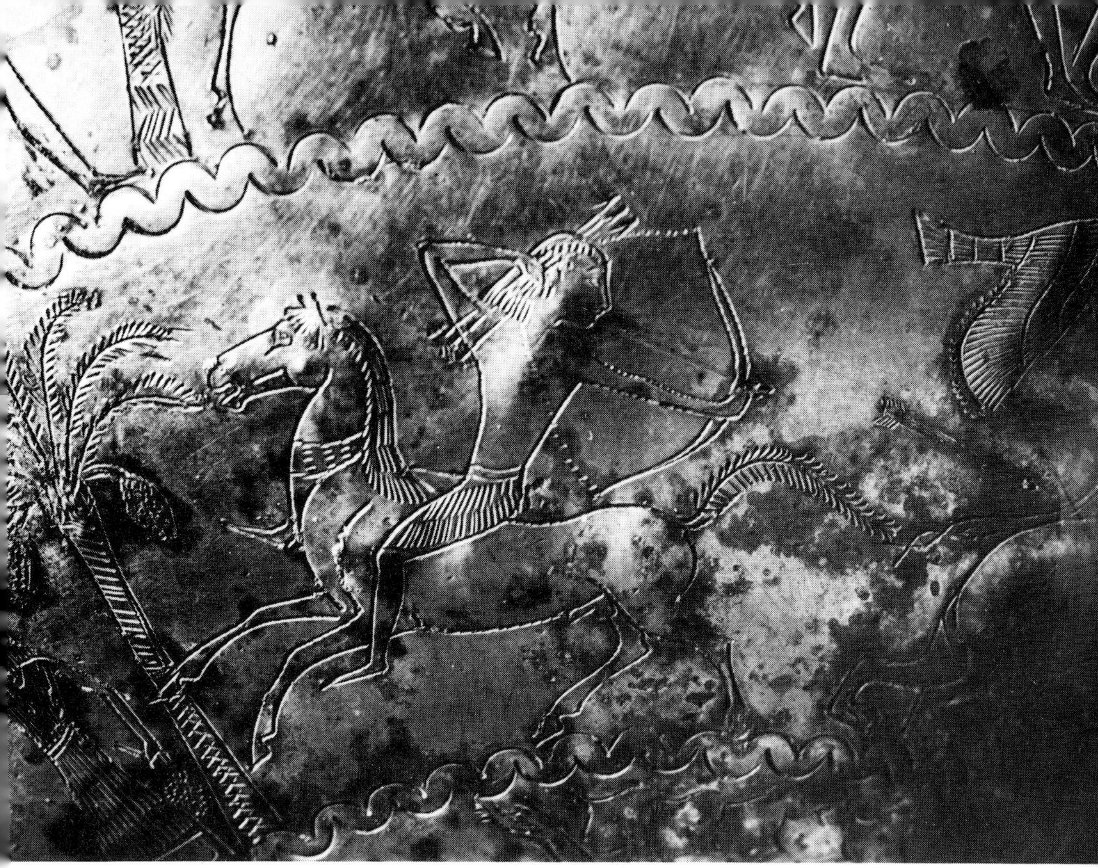

45. Etruscan silver bowl, seventh century B.C., showing mounted archer

46. *(below left)* From Metapont, southern Italy—mid sixth century B.C.

47. *(below right)* Etruscan vase, sixth century B.C.

48. This motive from the neck of the Vix Vase shows a war chariot with four horses, *c.* 500 B.C. The vase was taken from the grave of a princess at Vix, Châtillon sur Seine, France

49. A silver cult bowl from Gundestrup moor, Jutland, possibly first century B.C. The riders' spurs and horse harness are north European

pooled their resources to own one mare.

The story is different with the sheiks and rulers. Many owned large numbers of mares and saddle horses.

There is evidence of both riding and driving horses in Europe from *c.* 2000 B.C., but since horses were not required for military purposes there were apparently no marked advances for many centuries either in breeding or training. The Europeans of those days went about their daily business without any obvious desire to expand their territory at the expense of other peoples. The rich grasslands of Europe were probably more than sufficient for the needs of their herds.

Right across Europe there was, too, the very active early Iron Age Hallstatt civilization and culture. The Hallstatt period received its name from the huge burial ground which was discovered near Hallstatt in the Salzkammergut, Austria. Throughout the period there was considerable trading in gold, bronze and animal plastics made by local artists. There have been recovered many models of horses with riders and two- and four-wheeled carts. One small lead model has a team of twelve horses harnessed four abreast, quadriga fashion. These came from the burial mound at Frög, near Rosegg.

There is, too, a bronze sceptre with an elegant head of a horse, and the famous bronze situla found at Vače, Krain, in Yugoslavia. Thus, from Dümmersee *c.* 2000 B.C. down through the centuries useful, if unassuming, horses were being bred by people who used them simply for transport, for pleasure (hunting), or for religious purposes. These inhabitants of Europe would seem to have been real horse lovers, even although we know little about them. They appear to have belonged to a widespread class of peasant farmers and traders.

The silver and copper bowl found at Gundestrup is probably Celtic in that the idea of a stag-god with horns is of Celtic origin. It is a northern Gallic motif of the first millennium B.C. And in trying to discover the origin of the equestrian Celts—who appear geographically at Hundesingen (Singing of Dogs), a place situated almost at the point where the source of the Danube meets the source of the Neckar, the question arises if there is here possibly a connection with the People of the Deer—the Saki of Asia?

Strabo tells us of a northern nation of Scythians called Celto-Scythae. Like the Scythian tribes, the Celts used an Indo-European language which has been, unfortunately, little translated into any modern European language. If it had been we might learn something of their knowledge of horses since they had a natural flair for horsemanship and they seem, like their Asiatic contemporaries, to have had an upper class of warrior nobles.

The Celts were closely connected with the ancient world but appear to have been independent of native tradition and they brought a new realism into their art and their mode of living. They traded throughout Europe and during the succeeding centuries established themselves in Spain, France and the British Isles and, of course, Central Europe.

If we compare ourselves to the Belgae and the fourteenth-century inhabitants of Britain to the Bronze Age Celts, then it becomes clear that we have absorbed and improved upon Celtic transport, farming, language and culture. Just as the fourteenth-century inhabitants of Britain were cultured people, in many ways some may think more cultured than today's civilizations and in other ways more primitive, so the Celts could not have appeared to Julius Caesar wearing woad and skins, as some history books suggest. In the light of our knowledge of early Middle Eastern civilizations—of which Caesar knew nothing—this suggestion is ridiculous. The men of woad and skins had fashioned flint heads in Norfolk, Suffolk, Wiltshire and Oxfordshire in Palaeolithic times (i.e. some 50,000 years ago!). Readers would do well to consider the advances of Stone Age man, if they wish to visualize in its proper perspective the 'modern' advent of the Celts into the British Isles and the even more 'recent' advent of the Belgic arrivals who preceded Caesar by less than a century and who were, in fact, still coming in, just as the Boers and the Bantu simultaneously invaded South Africa.

The same method of consideration must be applied to the movement of horses or ponies into Britain. The earliest undoubtedly *walked* here, with or without the assistance of Stone Age man. Why they came does not concern us; we know that they were the prey of the hunter. Some were a Tarpan offshoot and developed into the Exmoor pony, others may have been of the *Equus robustus* strain.

It is possible that the Ardennais strain of the day found its way into the hands of the Frisii and that the Belgae, having suitable boats to carry horses, brought either the early pre-Friesian horse with them or the Ardennais (*Equus germanicus* of some writers)—and the ugly German horse of Caesar's time, or, as Tacitus said in the first century A.D., "enviable neither for beauty nor for fleetness". But perhaps they would have carried an armed Belgic warrior better than the Exmoor pony strain.[25]

Although the Belgae warred with the Celtic Goidhals, they were rapidly absorbed into Celtic civilization which they enhanced by their vigorous methods of agriculture and politics.

We are thus able to see the pattern of Celtic equestrianism: chariots drawn by a pair of ponies *c.* 750 B.C., and Belgae riders of 75 B.C. In the Ireland of A.D. 450 chariots were still in use. There are chariot burials of both the Dumnonii and of the Parisii. Long before Caesar, both chariotry and riding were common ways of travelling throughout Northern Europe.

[25] The earlier Celts had worn bronze helmets, scale-mail, sheet bronze or leather cuirasses, and carried shields, swords, and daggers.

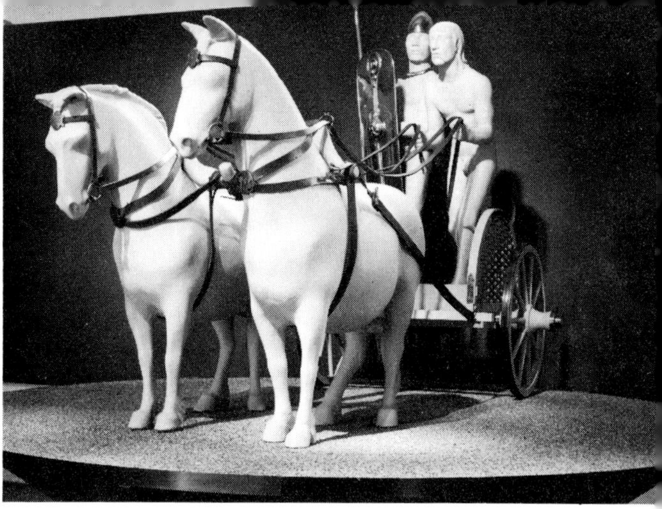

50. *(above left)* Celtic horse armour: the Torrs chamfrain

51. *(above right)* Celtic chariot

52. Roman warrior

The strain of *Equus europaeus* seems to have been spread across Europe, and at this point, except as I have already mentioned, we hear and indeed see little of *Equus robustus*. Perhaps we may infer that this type was less common, as indeed it is today, but still it did exist, as we see from the model of the bronze group of the goddess Epona riding a cob mare with foal at foot which was found at Alise Ste Reine.

The Celtic tribe of the Parisii produced charioteers of fame and they, too, belonged to the group of Brythonic peoples who arrived and settled in Britain. Other Celtic tribes who came over from the Continent were the Brigantes, who settled in Spain, southern Ireland and Yorkshire and Lancashire (they were good horsemen); the Trinovantes of Essex and Suffolk; and the Iceni of Norfolk and Suffolk. The Iceni especially were a horse-breeding tribe having a warrior class of charioteers. Their two-horse chariots repelled the first Roman invasion of Britain and after the defeat of their Queen Boudicca, Iceni chiefs were ordered by the Roman General Severus to return home and continue with their horse breeding and to stop meddling in affairs of state! The strange fact is that the Iceni centre of horse breeding was about thirty miles from Newmarket in Suffolk, which is situated near the long barrow presently being excavated, and which some 2,000 years later became the first most famous centre in the world for breeding the Thoroughbred horse. The breed for which the Iceni were famous was *Equus europaeus* which the Exmoor pony closely resembles (illustration 7).

The Coritani who settled that part of Britain now called Lancashire were a quiet agricultural people, and since most settled farming communities were animal breeders, we may suppose that they, too, bred horses. The Celts had a variety of chariots, traps, gigs and other horse-drawn vehicles,[26] whilst at Newstead under the Roman occupation, remains of native ponies and a 14 h.h. better-bred horse which had probably belonged to a cavalry soldier were found. The remains of a few larger big-headed animals were also present and these may well have been imported from Europe.

From Caesar we learn that the pair of small horses yoked to the British war chariots was so active and well trained that the horses could be checked and turned in a narrow space or pulled up when going at full gallop on a steep declivity, or made to stand whilst the warrior ran out on the pole and stood on the yoke. From what Caesar says of the horses of southern Britain and from statements of Dio Cassius, the horses of the Caledonians up to the end of the first century A.D. belonged to the fine-limbed 'Celtic' type or to a cross of the 'Celtic' and forest types, under 12 h.h. The 14 h.h. horses closely resembled the modern Arab. Some horses may have measured nearly 15 h.h. but the majority were below 14 h.h. rather than above this height.

[26] See Arthur Weigall, *Wanderings in Roman Britain*.

From the second century A.D. onwards, the Gauls imported well-bred stallions at great cost from the south of Europe, to improve their strains; and horses belonging to cavalry and mounted infantry may have originally come from Gaul.[27] The Gauls themselves were keen harrier hunters. In Gaul the war chariot was made redundant about 120 B.C.[28] In Scotland, war chariots are mentioned at the Battle of Mons Graupius and even as late as Septimus Severus's expedition in A.D. 207.

The bronze chamfrain found at Torrs is of the La Tène period and reminds us that horses were accoutred with armour as in Hittite times.

Information on horse breeding during the period of the Roman Empire is scattered. Spain was a leading province and horses for army purposes were especially bred here. Pliny the Elder (A.D. 23–79), who was at one time Procurator in Spain, refers to horse breeding: "The Galician and Asturian tribes of the north breed a kind of horse which they call thieldones or celdones. This smaller strain which we call Asturcones, do not trot but have a special easy pace produced by alternately moving both legs on the one side." Pliny in *Historia Naturalis* clearly refers to the amble or pace. In fact, a very great number of cavalry horses were bred in the different Roman provinces of Europe. There was a squadron of Frisian cavalry stationed on Hadrian's Wall, where ten different horsed units have been named between it and the Antonine Wall further north. These also included the Hungarian First Pannonians and the Scythian Sarmatians and a cohort of Syrian archers. I wonder if it was from these same Syrian archers that the inhabitants, both Roman and British, of these islands first learnt about mounted horse sports and racing? Racing became a sport under Roman rule and the archers (descendants of the Hittites) would have brought well-bred eastern horses with them. We know, too, that the Romans obtained heavy horses from the province of Noricum. The Noriker today is a light vanner/cob and it is unlikely that it was ever any heavier, but compared with the eastern horse it would appear to be 'heavy'.[29]

They crossed these horses with their own lighter strains in order to obtain cavalry horses capable of greater endurance. On the Capitol in Rome there is a statue of the Emperor Marcus Aurelius dating from the second century A.D. which shows the result of the cross fairly clearly. Roman emperors were not renowned for humility and no doubt the size of the horse has been exaggerated. In the occupied provinces, at any rate, Roman rulers are mounted on less massive animals.

[27] For a survey of Celtic religion and worship of the horse see *The Foals of Epona* by Daphne Machin Goodall and A. A. Dent.

[28] *Pre-history of Scotland*, by V. G. Childe.

[29] See Daphne Machin Goodall, *Horses of the World*.

In excavations at Hadrian's Wall, Professor Ewart discovered the bones of six different types of horses and ponies and some intermediate types which were possibly cross-bred animals. It would seem then that by this time—*c.* second century A.D.—*definite* breeds had evolved as opposed to *strains*, which appears to me to be a better description until we reach this period. Ewart (and later writers) confuse the issue by calling a certain 11–12 h.h. pony with a slightly dished face "the Celtic pony". There was never a "Celtic" breed but a strain *Equus europaeus* which flourished in Celtic times and was much used then. I do not think we made the position much clearer when we described the same strain as "North Atlantic" in *The Foals of Epona.* If one could study Gaelic or Ibero-Celtic, there probably exists a proper term for these ponies. The description, however, does more or less fit the Tarpan.

By this time there was a recognizable 11 h.h. Shetland pony—not surprising when we recall the cave painting from Lascaux where five types are depicted.

There was also a fine-boned animal similar to the horses used by the Hittites (and perhaps used, in fact, by a Syrian archer!) but somewhat larger, standing about 14 h.h. There was also an Exmoor type of 12–13 h.h.—the Exmoor did, of course, exist *as such* at this time and earlier. The *forest* type was also present, thickset and with a longer back—the kobyla of eastern Europe or other mountainous regions, perhaps. And lastly, a 14–15 h.h. much heavier-limbed animal with convex or straight face, apparently designated by the experts of the time as the Siwalik type from the range of hills in northern India where similar fossils are said to have been found. Again, such a designation adds nothing to the study of horse breeding. We know these animals could not have come from India. The heavy horse of the diluvial period has been found in Europe and it may over hundreds of years have 'improved' into the useful cob—or degenerated into a locally heavy-boned animal. This could happen in a matter even of a few generations by injudicious breeding and feeding. And in addition, German cavalry units from the Rhineland probably did bring over heavier animals.

There were at least two British cavalry regiments, the Ala Prima Flavia Augusta Britannica stationed in Hungary and the Ala Quinta Brittonum in Morocco. It was a rule or general practice of the Romans not to keep native soldiers in their own country. Unfortunately there seems to be no information as to whether they were horsed on home-bred animals—which is more than likely—or on those of the country where they were serving—which seems less likely.

There *were* equestrian tribes amongst the Celts in Roman Britain, as I have mentioned earlier, and amongst them were the Brigantes, Parisii and Votadini. The Celtic civilization did not die out just because Romans occupied this island. Thus young men from these tribes will have formed these cavalry units and probably have taken their own locally-bred horses abroad with them, just as did the knights of the Middle Ages some thousand years later.

When the Romans withdrew from Britain during the fourth and fifth centuries A.D., King Cunedda of the Votadini, whose country lay in modern Lothian and Northumberland, led his people from the devastated north to the west, through the country occupied by the Brigantes (later the Dales and Fells), driving out Irish invaders who had slain many of the Ordovices of North Wales and taking over their lands. Taliesin, the bard, sang of him: "He would freely distribute war-horses in winter", thus helping mountain farmers. In this manner there probably came to Wales the heavier ancestors of the famed Welsh cob. There is a Roman road from Chester which passes Lake Bala in which locality the Romans are said to have founded a stud to breed pack-ponies. Lake Bala lies well within the country of the Ordovices, who had both lead and copper mines in their country. No doubt the pack ponies were needed to transport these materials. In later centuries the name Powys was given to this country and studs for breeding the Powys cob were also founded.

In an earlier chapter, we left the Chinese on the point of sending forth an embassy to obtain horses from beyond the Tien Shan Mountains (see Chapter 3), and later we discussed the quality heavier horse from the Plains of Nisea (page 114). It is, perhaps, a far cry to seek here a connection with the heavy draught breeds of Europe in the eighteenth and nineteenth centuries—and yet I believe there might well have been such a connection.

But first let us bring China up to date. In the last century of the first millennium B.C. the horses of Ferghana had begun to acquire a fame equal to that of the Nisean with whom they were in all probability closely related. It was recognized that cavalry horses which were descended from the Tarpan were better qualified for military purposes than the descendants of the Przevalsky horse. Joseph Weisner says that there is a stele in Shantung which shows both strains together.

The early historic and ethnic names remain geographically the same for many centuries. Now the great horse-breeding country Bactria, within whose boundaries the fertile valleys of Ferghana lay, occupied a very large area on both sides of the River Amu-Darya. This is now part of southern central Asia and north Afghanistan. We find a connection between Chinese activities of the second century B.C. and the Nisaean horse of Persia with the later heavier horses in Margiana, Parthyène (corresponding to present-day south Turkmen and part of Khurasam in north-east Iran), Chorezem (on the lower Amu-Darja) and Sogd or Sogdiana (the valley of the River Zeravshan). These regions were occupied by farming and stock-raising peoples and all the areas mentioned were famous for breeding superior horses: Ashkabad, Merv, Balch, etc. The

53. *(above left)* Chinese horse of the Wei Dynasty has bells across his chest. It is an example of the "grete hoked head" described by Thomas Blundeville writing in the time of Elizabeth I

54. *(above right)* T'ang Dynasty, Chinese horse, A.D. 618–906

55. Chinese Wei Dynasty horse, A.D. 524—the *Himmelspferde* or Heavenly Horses referred to by B. Melchers

Paricani, in particular, were a tribe identified with Ferghana.

To the west of the River Ili there was a tribe named Spotted Horses. They were "a large nomad tribe bordering on and resembling, but speaking a quite different language from the Kirghiz, living largely on cheese and kumiss and so called from the fact that their horses were mostly piebald. The Turks are said to call a piebald horse *ghora*; hence, this people are also called Ghora" (Parker). Their neighbours were a shy people who used the elk or moose-deer.

As we know, it was to the people of the Ili valley that the Chinese first sent for horses and quite possibly it was from this region that the spotted horse went first to the east and later through the early Crusades west, until finally, via Spain and the *conquistadores*, they became an accepted part of the equine population of the New World; in particular of the Nez Percé Indians who also many centuries before had themselves emigrated from Asia!

In 121 B.C. Chang Ch'ien, the envoy of the Chinese Emperor Wu-ti, was sent on his second mission and visited the Wu-sun who occupied the Ili valley, south of Lake Balkhash. As a result of these visits, the Wu-sun twice presented the imperial court of China with large studs of horses which were bigger but finer boned than the native Chinese horses. Wu-ti sent between five and ten trade missions annually to the states of Aorsi, Tayüan, the Yueh-chih and Parthia. Chinese silk, jade and coral were exchanged for horses.

However towards the end of the second century B.C., the ruler of Ferghana refused to barter his thoroughbred horses and murdered the Chinese envoys. At this, Wu-ti sent two large expeditionary forces to obtain the horses by force. Half the expedition died of hunger, thirst and exhaustion on the 3,000 kilometre march through mountains and deserts. But over 3,000 horses were collected. Some 2,000 of this number died on the long desert road back to China. Only 50 'blood-sweaters' and around 1,000 inferior horses and ponies survived.

The Ferghana horses were much stronger than the Ili horses and Yetts[30] says they were very similar to the classical Greek horses. Fernald[31] suggests that the Ili horses were cross-bred Ferghana × Mongolian pony. This is possible, for bas-reliefs show a powerful common horse with a convex profile. It seems that the Sien-pi Empire lasted until A.D. 200 and then the Hung-nu name almost disappeared. The Huns went westwards, appearing in Europe in A.D. 300 where they remained for about a century before returning to Asia where they merged with the western Turks. The Sien-pi had faster horses than the Hung-nu!

In his book *Unknown Mongolia* D. Carruthers wrote:

[30] Perceval Yetts, *The Horse: A Factor in Early Chinese History* (Eurasia Septrentalis Antigua, Helsinki 1934).

[31] Helen E. Fernald, *Chinese Art and the Wu-sun Horse*.

> About the third century B.C. there emerged into prominence a people who were destined to leave their mark on the whole civilized world. From the far south of Mongolia, perhaps from the borders of China (Shensi and Kansu) came a wandering people the Uigurs. Their origin is problematic. These tribes wandered north and eventually settled in Mongolia. Eventually spreading over the Yenisei regions as far north as the Chulin river, until in the eighth century A.D. their kingdom reached over the whole of northern Mongolia from Lake Kossogol to the Black Irtish . . . In the ancient Uigurs we have the origin of the Turkish race who later on overflowed all central Asia and made an empire on the shores of the Bosphorus.

Kankly is Turkish for 'a cart'. Parker says that the cart-using Hung-nu and Turks north of the desert, Issekue, and Syr-Darja or Upper Jaxartes were in ancient times called Ting-ling, then High Carts, then Terek and finally Ouigurs and Kankalis. So that these people (*c.* A.D. 600) were the remnants of the Huing-nu or Hung-nu, after they had moved westwards. "They were strong daring nomads, great horse-archers, and fond of raiding; as soon as the quite new name Turk became magical under the Assena family, they formed part of the Turkish dominion." (Parker) Later they formed their own dominion. The Kirghiz or Kerkur have lived in their present place for over 2,000 years; they are possibly related to the Kankalis or High Carts, possibly they may be the ancient Awsen (Ili?) or "nomads of Kuldja" (Parker). They tend their horses until full grown, when the best fighter is made leader of the herd. *They also possessed wild horses.* They lived on kumiss and their language is similar to the Ouigurs.

Between the T'ao and Yellow Rivers, just east of Koko-nor (A.D. 428) The Tukuhun founded a dominion. It was a bleak country. There was a small island in the centre of Koko-nor and every year when the lake was frozen a number of fine mares were driven on to the island; the foals were collected the following winter. A number of splendid Persian mares were obtained by the Tukuhun for this purpose and their young obtained great repute for swiftness as 'Koko-nor' colts. Parker also tells us that when the Chinese made a treaty with the Hung-nu *c.* 48 B.C. a white horse was slaughtered. The Tungusic rulers of China often sacrificed white horses in the fifth century A.D.; whilst Genghis Khan used the skull, encased in silver, of the Khan of the Keraits and ratified his first military league by the sacrifice of a white horse.

We come to the representations on bas-reliefs of obviously different strains of horses which cover a period of about a thousand years. A stone rubbing of the Han Dynasty[32] dated A.D. 25–220 reproduces an early representation of one of the famous horses of Ferghana—or an early descendant bred in China. It shows a massive, well-proportioned body with

[32] This dynasty was founded by Liu Pang in 202 B.C. when China extended her empire after combating the Hun tribes.

good bone, a small head set on a strong neck and shoulders, and a very broad chest. The horse is bridled and saddled, with a point of great interest—there are stirrups attached properly to the saddle.[33] The horse is held by a booted soldier! Another bas-relief rubbing is of a fleeing horse, in many respects similar, yet it reminds one more of a large steppe pony with a flowing mane rather than a Ferghana horse. It, too, carries a saddle with stirrup. Then in the Wei Dynasty, A.D. 220–1279, we have two pictures of Mongolian horsemen. One is mounted on a spirited steppe pony and is chasing a wild horse (or one belonging to a tabun). He has a sling to catch it with. These slings on the end of a pole are still used for the same purpose today. The second representation shows a Mongolian with a large mettlesome horse on the longe. Another bas-relief is of a pacing parade horse in all the glory of a Spanish descendant of 1,500 years later—the rider is apparently riding sideways too!

Throughout the T'ang period China bred some first-class 'great' horses in all respects similar to those Nisean horses bred at the same time by the Medes and Persians. We have many representations of them. As well as these very substantial cavalry horses, there is a most interesting rubbing of two women mounted on Achal Teké horses, a black and a chestnut; the small head carried high, large eye, short back and long legs are unmistakable.

Around A.D. 1000, the Nuchen (Manchu) who resided in the Sungari River valley and the lower reaches of the Heilung Kiang River, had to deliver horses to the Liao Dynasty (907–1125) established on the upper reaches of the Liao River and part of Hopeh Province.

A stone rubbing of the Mongol Yüan Dynasty (*c.* 1277–1367) shows a groom with a cob-size, bridled, therefore domesticated, mare. The mare has been brought to a dun steppe stallion, although this may be coincidence of colour chosen by the modern colourist. The practice of mating domestic mares with wild or semi-wild stallions had been carried on by the steppe nomads from earliest times. Indeed, it is thought that the first mules were bred in this way, by tethering onager mares in the neighbourhood of a wild stallion.

Around A.D. 1008 we have the Cathayan Empire of northern China and the Sung Dynasty of central and southern China. The Chinese said of the Mongols, Turks and Hung-nu, Manchus, Cathayans and Sien-pi in each case "horse's back forms the state". The horses of the Ghei were very hardy and active and about A.D. 842 they were found westwards in the region of modern Kalgan. Ghei and Cathayans are properly the ancestors of Mongols, occupying their old quarters. And Eleuths or West Mongols are remnants of old Hung-nu and Turkish Empires.

"The same 'horseback' hordes have from time immemorial assailed China, Persia, Mesopotamia, Syria, Babylonia and even Egypt, always

[33] The stirrup is supposed to have first appeared some 200–300 years later!

introducing a new 'horse' era." (Parker). The chief occupation of the Cathayans was horse-breeding and in order to conquer them, the Chinese set fire to the prairie grass so that the horses were either starved or destroyed. In A.D. 1003 the Cathayans had 18,000 Chinese cavalry at Peking. The armies of their own tribes and the four chief garrisons north of the Wall numbered about 180,000 horsemen. Of these, 5,600 were the monarch's personal guard and 94,000 were raiding troops.

There is a further note that about 986, when Wei-chang was on hunting expeditions, wild horses were met with.

We have seen that the idea of knighthood originated in ancient Asia when war-chariots were used. The idea of mounted knights on 'great' horses stemmed from the Persians who also wore armour, although the Samartans also covered themselves and their horses in leather-scaled armour. Slowly the idea of knightly cavalry spread westwards in the first millennium A.D. It was an outlet for younger sons of noble families who joined religious orders devoted to combating the 'heathen' whether they were the autonomous inhabitants of the Baltic lands or the non-Christian occupants of Near-Eastern countries. Although the Crusades or Wars of the Cross ultimately failed in their spiritual objectives, they benefited Europe indirectly during the 300 years. Trade was considerably stimulated between Europe and Asia. From 1187 to 1291 Acre became the capital of Christianity. Shipping increased for the transport of armies and for commodities from the east.

North Syria, always noted as a horse-breeding country lay, of course, directly between Persia and the west. Since we do know that returning Crusaders in many instances brought unspecified oriental horses with them, it is extremely probable that the Nisean was amongst them—perhaps not in great numbers but sufficient to influence studs in Italy and Spain. The "hoked head bending like a hawk's beak" was known in the East, and there is every reason to suppose that it reached the West. According to Marcus Fugger it did! As we have seen, the convex profile, a markedly distinguishing characteristic in any horse, had reached China and similar horses with similar characteristics could also have reached Syria during the twelfth and thirteenth centuries—or earlier.

The Indo-European group of the Germanen who were to be found from the Rhine eastwards to the Vistula were co-existent with the Celts and Romans but obviously living outside the Roman Empire. They were the immediate followers and probably the descendants of the Scythian tribes who had penetrated Europe, the Lausitzer civilization and the Late Bronze Age peoples of the north.

There were three groups: those of Norway, Sweden, Denmark and Iceland; the Saxons, Franks, Alemanen and Bavarians; the Vandals, Burgundians and Goths, and during the second century B.C. the Cimbrians and Teutons who from the west coast of Schleswig-Holstein together wandered to Italy.

At this period, because of climatic changes in Asia and Europe, a mass migration (the *Völkerwanderung*) of peoples took place. Such a disturbance of populations was not to be repeated until the wars of the twentieth century.

About 150 B.C. many upheavals occurred which influenced the dislocation or even the complete destruction of studs and breeding of horses. Apparently the climate of Mongolia deteriorated rapidly and the inhabitants had to move into a milder climate for the sake of their flocks and herds. They drifted to the east, south and west, found better conditions and prospered. Their herds and their horses improved and they settled in the occupied plains and valleys of civilization.

In the west further events were triggered off. The climate of Europe was also getting worse. Scandians set out eastwards. The Asiatic Huns began *their* drive, terrorizing the Slavs in European Russia and the Chazars who then inhabited the Caucasus. Their dismay was communicated to the Germanen and the Armenians who in turn moved away from the advancing Golden Horde in A.D. 375, who seventy-five years later settled in Austria and Hungary.

Finally, Mohammed in Arabia launched a further movement and the Saracens invaded the remainder of ancient Europe and the Near East.

It was general amongst nomad steppe peoples to take over the name of the greater conquering tribe. Thus the powerful Avars included many peoples under this name. But just at the time when mounted hordes from Baian appeared at the Pontus, no single nomad people wanted to boast of their connection with the name Avar, whose kingdom at the time lay under Turkish dominion. These Avars are, in fact, identical with the Zoan-Zoan of the Chinese and held the balance of power in the nomad world.[34] When their kingdom was destroyed, they fled to the Chinese, apparently to the Tungrish Mukri people. The Avars who subsequently settled in Hungary, according to the Turks, were described to the Byzantine people as runaway slaves. In 563, after confused wanderings, they arrived amongst the Alanen of Caucasia where they were quite unknown, and when their first envoys to Constantinople appeared, "the entire town gathered to look at them, since no one had ever cast eyes on this people". "This people" resembled other nomad riders except for their long pigtails decorated with cords. John of Ephesus mentions "refugee nomadic hordes with braided hair which were called Avar after the fashion of their hair", and the Chinese mention the Zoan-Zoan who had braided hair and were therefore called "Ropebarbarians" or "Rope-

[34] Alföldi, "Zur His. Bestimmung der Avarenfunde", *Eur. Septent. Antigua*, 9–10, 1934.

headed barbarians".[35]

Thus both branches, the Hungarian and the eastern Avars, are identical. Their tribal name was, in fact, Varchoni (Alföldi) even though the Turks described them collectively as Huns or Oguz. J. Németh draws attention to the Mongolian tribal name *Varguni.* Theophylaktos lumps them together, the Avars from the Caucasus and the refugee Turkish Byzantine Avars, as "Uar" and "Chunni". And this name has been repeatedly used by many later writers although it is not entirely correct. However, it is probably the origin of the word *Hun* and *Gar* = Hungary and *Bavar* = *Bajovaria* = Bavaria, just as the origin of the later fabulous herds of horses upon the plains of Hungary was the steppe Tarpan.

And behind the upheaval of mankind, the earth continued disturbed. "Intense cold descended upon northern Euroasia, crops failed in central Europe and arroys incised the barren lands. Roman armies disbanded or were annihilated. Rich forests turned to scrub, or faded into bare gravel and sand and great rivers vanished," (Sanderson).

By A.D. 250, the mass migration was in full swing. The West Goths drove through the Balkans and Italy, southern France and northern Spain; the East Goths moved to Italy; the Vandals to Africa; the Alemannen, Burgundians and Franks crossed the Rhine; the Angles and Saxons invaded Britain; the Langobards moved into Italy. The old order vanished and the foundations of modern Europe were laid. This period also saw the beginning of Russia. Obviously breeds and strains of horses were mixed to a degree so that a 'pure' breed of any description can scarcely be contemplated.

The herds of European wild horses were decimated. Yet here and there local 'pockets' of ponies peculiar to an area still survived. They remained in their isolation, more or less undisturbed for centuries and probably some of them, at any rate, had very little influence on any subsequent evolution of the species *Equus*. In Europe some, however, did help to found warm-blood breeds. But herds like the Sorraia of Spain, the Tarpan of eastern Europe, the Exmoor of Britain, and of course the most primitive of them all, the Asiatic wild horse of Mongolia, remained, as they do today, within their areas of habitation.

In the British Isles, between A.D. 500 and 900, horses of several different kinds continued to be bred and sportsmen indulged their love of hunting stag and hind, as illustrations on gravestones in the churchyards at St Andrews and Aberlemno, Scotland, show.

The Tchertomlyk vase in the Hermitage shows a stirrup formed by a strap in the shape of a loop coming from the girth. The vase was found near Novo-Alexandrovka, Melitopol (Crimea) by W. W. Arendt,

[35] The Ugrian language group has three related languages which include Magyar (Hungarian) and the language of two small Siberian peoples, the Khanty and Mansi, and the Finnish group.

Moscow. The Avars are thought to have been the first people to use the stirrup in Europe,[36] and since it seems fairly evident that the Avars were descended from Scythian tribes, it could well be that it was from the Scythians that the Avars inherited the stirrup. The horses of the Scythians were fast, and it is difficult enough to ride a fast horse for long without stirrups without falling off; the obvious thing is to attach a loop to the girth.

There seems little reason to doubt Arendt's statement even if further proof is not available.

"In Assyria," says Arendt, "this problem [to retain balance] was resolved thus: a warrior mounted beside the bow man, conducted the fighter and covered him with his shield. The absence of support already impeded the nomad warrior, and thus the Scythians produced the first stirrup."

[36] Miklos Jankovich, *Pferde, Reiter Völkerstürme*, BLV.

5. The Cob and the Great Horse

Because the heavy horse breeds, which were selectively created in the eighteenth century, were rather unfortunately designated as 'cold blood' we have developed an erroneous and entirely imaginative impression of the war horse of the Middle Ages. We imagine that these war horses, the *dextrarius* of the knight, were massive 17 h.h. animals, rather like the Shire of today. We imagine that they must, of necessity, require *weight* to carry weight.

Nothing is further from the truth. The average knight was a small man, not above 5 feet 6 inches and with an estimated weight of 224 kilogrammes (about 34 stones),[1] and no knight in full armour could ever have mounted a 17 h.h. horse. It is nearly impossible to mount such an animal from the ground, even when one weighs only 10 stones. We also forget that one of the greatest faults of the nineteenth-century heavy horse, indeed of almost any breed, was the tendency to soft bone and to side-bones.

I do not imagine that if such heavy horses had really existed, when knights rode forth to battle, the knight would have got far in destroying his enemy—mounted, as the Saracens were for instance, on fleet Turkish horses. The Shire horse type may be comfortable at the walk, but he would be impossible at the trot, and heaven help the knight if his steed broke into a canter. No, the Shire and his kind were bred on the principle of 'weight to *move* weight' and the latter-day colossus was more modish than useful.

So what did the knight ride? None other than the cob, which now had a chance to evolve and enhance the equestrian world. Had there existed larger horses during the reign of Charlemagne (A.D. 742–814), he would not have been portrayed in all his royal splendour mounted on this really first-class cob. There is nothing of the heavy cold-blood horse here, and yet, if the splendid Percheron or Boulonnais had existed, Charlemagne would surely have insisted upon riding the larger animal to display his might? But this is a real cob: small head, short neck, powerful front and short back, excellent bone, no feather and obviously active. *This* is the

[1] L. S. Ruhl of the Kurhessen Museum produces statistics on the average weight of a knight in armour and horse with armour.

56. *(left)* Charlemagne riding a useful-looking cob. Such horses were to be found throughout Europe at this time

57. *(above right)* Silver-gilt dish from Persia, eighth century A.D., showing a cob-type horse

58. *(below left)* A sixteenth-century knight (from a German painting)

59. *(below right)* A knight on the march riding a pacing stallion, sixteenth century, Marienburg

animal to which writers of the period refer as the cold-blood horse. There is very little reference to these horses except in plastic form and there is no reference at all to larger horses.

At exactly the same period, the eighth century, a Persian king rides his cob to hunt the lion, wild boar and bear, just as did his predecessors, the Sassinid kings, on lion hunts a few centuries earlier.

These cobs must have been fairly universal throughout Europe and those parts of Asia which traded with Europe. I suspect that they flourished wherever the food was good and sufficient, good grass or good clover. A strain had developed in the Ardennes, for Caesar had already found horses useful for military purposes here, just as Napoleon did many centuries later. The cob strain also developed in Normandy, in the mountainous areas of Austria and Norway, in East Anglia and Wales, and, indeed, in the Highlands as well.

A colossus is certainly not needed to carry weight. What is needed is a stout-boned, short-legged, well-bred horse with a short, strong back and plenty of muscle, good joints and feet, with a fine head and a responsive mouth. This description does not fit the European heavy horse breeds but it fits the old Suffolk cob, the ancestor of the Punch. It is unfortunate that the lighter-bred eastern horse, which has been crossbred into so many breeds, has almost completely ousted this hard and useful animal. But in 1767 Sir Richard Smyth wrote to George Selwyn, telling him that a certain Dr Thistlewaite is dead and that his horses are to be sold:

> Amongst them is a little bay gelding, about thirteen or fourteen hands, with flaring long tail, strong enough to carry you, the Mayor, and all the money you ever spent in elections at Gloucester together. The Doctor, some forty-eight stone, always shot off his back, and the keeper killed all the deer from him. I mention this as proof of his sedateness. He goes fast enough to carry you close to foxhounds in full chase; but if your affairs do not require so much attention a snail would distance him. His figure is such that if you were to meet a tailor on his back you would pull off your hat to him, though you did not owe him one shilling, I know twenty men of weight who want him, but the weight of metal will have him.[2]

The good doctor, therefore, rode 14 stones *more* than the knight in full armour of a few centuries earlier—and hunted his cob as well. In the eighteenth century, of course, hunting was a slower sport altogether, becoming a galloping sport in the nineteenth and twentieth centuries when blood horses were introduced.

A. Hink, writing in 1909, says: "The idea that the larger domesticated animals came from the East becomes more and more likely."[3] Indeed, the

[2] Lord Ribbesdale, *The Queen's Hounds.*

[3] A. Hink, *Über schwere Pferde im alt- und mittel-persischen Reich*, 1909, Zeit. f. Gestütskunde u. Pferdezucht.

comparatively heavier horses depicted on the Persepolis would appear to form a basis for this supposition.

As we have already seen, Urartu already bred a more 'solid' strain, whilst some centuries later the chalky plains of Nisea produced horses of a heavier calibre than the usual eastern horse, the smaller being the cob. Ferghana also produced heavier horses and the mural of the seventh century A.D. shows us a real heavyweight horse. The blood horses of the T'ang Dynasty also had their origin in Ferghana.

Herodotus describes the famous royal stud on the Nisaean Plain in Media, where the best horses in the world were bred—as many as 150,000. "The horses are called Nisaean and this Plain produces the largest horses." No doubt this was due to the chalky soil which helps to form good bone (as did the soil of Ireland a thousand years later!). The Persians were fine men of Aryan type whilst the kings and aristocracy were tall. And they could not mount their larger horses as the Greeks did, by simply springing on to the horses' backs. Since the Persians had neither stirrup nor saddle they helped each other to mount and the method of mounting was 'in the Persian style'.

Returning to Herodotus, we learn that Cappadocia had to pay a tribute of 1,500 horses and 2,000 mules, Media double that number, and Cilicia 360 white horses.

"Here," writes Nearchos, referring to the mountain plains of the Zagros Mountains,

> the climate is mild; the country is rich in plants and well-watered meadows and all other fruits except for olive trees. There are beautiful pleasure gardens, rivers with clear water and lakes that are rich in river and sea birds, water the countryside. Horse-breeding flourishes and that of pack animals as well.

Hink estimates the height of the Persian horses as around 17 h.h. He also says that they were probably bred in Parthia, Bactria and Sogdiana. The war horses of the Massagetae, which were also fairly heavy, wore copper breast armour.

The Sassinids of 226–636 A.D., who later occupied the same territory as the Persians, much preferred "heavy cavalry". Like their predecessors, the Parthians, they used the saddle and stirrup. They left their light cavalry to their Arab and Armenian confederates. But their own horses were big and strong.

There is thus every indication that these quality heavier horses were bred in the lands of the ancient Persians, Parthians and Sassinidae, and that they were brought to Europe.

I personally think that the East European *Equus caballus robustus* (kobyla) wandered or was taken from Europe eastwards, perhaps via Cucuteni 10,000–2000 B.C. (Tripolje civilization), Maikop 2000–750

60. Bamberg rider, from the cathedral, an example of the 'Great Horse'

61. A knight mounted on a light, possibly spotted, eastern-type horse. From Baldishol, Norway, twelfth century A.D.

B.C., and across the Caucasus to Urartu and Media. During many generations the strain was consistently improved by expert selective breeding—the durations and stability of kingdoms as well as the lives of rulers depended upon good horses. This blood eventually returned to Europe, where, during the Middle Ages, it fused with the European cob strain, also with that of the imported *Equus orientalis* and with *Equus europaeus*.

It seems perfectly feasible that returning Crusaders who brought Eastern horses back with them should also have brought a number of the heavier horses, which were possibly of Nisaean stock. It seems difficult otherwise to explain why the term 'Great Horse' was used *only* when referring to the horses of the knights. It follows that these, when crossed with the cob, which, as we have seen, was present in many parts of Europe, would produce a heavier strain of horses than had previously been seen in Europe. Coupled with this, agriculture was well established and better foodstuffs were available for both human and animal populations.

Again I repeat, what we call *heavy* in the twentieth century did not exist as a breed anywhere previous to the late Middle Ages or early Victorian expansion of trade and agriculture when weight was needed to move weight. I make no apology for stressing this point. It all depends upon what we envisage by 'heavy' and many of us have grown up with a false image.

Paolo Uccello's "Rout of San Romano", 1456–60, gives us a fair idea of the size of the Great Horses of the knights and the strain does not appear to differ much from the mount of Marcus Aurelius, and, as a point of interest, they have convex faces. Gattemelata (1450) also rides a horse of the same strain, as does Dürer's knight in 1513.

In England by the sixteenth century however, Thomas Blundeville, writing in the reign of Elizabeth I in *The Fower Chiefest Offyces of Horsemanshippe* refers to breeding horses for service "whereof this realme of all others at this instant hath greatest need". Blundeville had become concerned at the shortage of good horses and felt very strongly that by increasing the breeding of horses an attack from France might be prevented. After all, the Norman Conquest had happened only a few centuries earlier and English cavalry had been in no shape to withstand the onslaught of Duke William's knights.

Besides his hackney or easy horse to ride, a gentleman should also keep a horse "mete for service in the field". From Blundeville's account there was a fair choice of horses for the gentleman of the sixteenth century. These included the Turky horse, horse of Barbary, Sardinia and Corsica. The Neapolitan or Courser of Naples, Jennet of Spayne, High Almayne, Flanders horse,[4] Frizelande horse, Swiethlande horse, Horse of Hun-

[4] A horse of the province of Flanders, rather than a breed.

gary, Iryshe Hobbye.[5] These were the breeds that were brought to Europe or were bred here. (I have retained the original spelling.)

Blundeville goes on to discuss the Eastern breeds, such as the Parthians, which, citing Vegetius, he says were big, bold, courageous, well-trained, easy ambling paced horses, but the horses of the Medes had better conformation (Nisaean!). The Cappadocian horses were, he says, descended from the Parthians but had heavy heads. Greek horses possessed good legs, great bodies, comely heads and were of high stature. He commends both the Pellians and the Magnesians but thought nothing of Thracian horses. And he comments that the Turks neither amble, racke nor trotte but have an easy trayne". All the above breeds were imported into England. The Barbers (Berbers) of Mauritania were especially recommended. The Libyans were stronger made, longer bodied, thicker ribbed and very hard, and the mares liked music. The Cirenians[6] were big horses with small gaunt bellies, they were fast and good workers and were valued in ancient times for chariot racing. The "Barbs of the king of Tunnys [Tunis], Massilia and Numidia are small, fast and go long distances, we esteme them so much. Vegetius says they come from Spayne". But the horses of Sardinia and Corsica were bad-tempered and inclined to rear.

Blundeville considered that the Courser of Naples excelled other breeds, they were trim horses, comely and strongly made, of so much goodness and of gentle nature and high courage. Excellently proportioned, long lean-headed, from the eyes down "bending like a hawkes beak", with a lofty pace, clean trotting and strong gallop and rein with better grace (more easily trained!). Even so they were "better in earlier times" (see illustration 53).

The Jennet of Spayne was finely made, of slender buttocks and the nobility liked them for speed; they were faster than the Parthian. A completely fixed breed but "they lack courage in that they carry their riders off the field" (in battle!). Of the Hungarian, he says, "a great hoked head with eyes outside, mane to knees, long bushy tail and his body is full of empty corners". They are long rather than high. The pace is a hard trot, he is temperate, wise and hard and can endure cold and hunger.

The Almaine (German) is commonly a great horse with a hard trot using his hind quarters but not his forehand and the Flanders, like the Almaine, is used for draught. The Frizelande is a mean stature horse, strong and compact, good legs, fast for carrière and ménage but of a devilish and stubborn nature; however he has a comely trot.

The Sweacian or Sweathland (Swiss) which is imported is commonly pied of two colours, wall-eyed and docile. The gentlemen of that country liked coloured horses. Of the Iryshe hobby, he says it is "a pretty fyne

[5] Which had Jennet (Ginete) blood and was valued for racing.

[6] Cyprus.

62. *(left)* Spanish war horses of the late eleventh century A.D.

63. *(right)* Horses depicted on a stone in the churchyard at Aberlemno, Scotland

horse", with a good head, good conformation, tender mouth, nimble, light, pleasant and "apt to be taught". They are amblers and swift but are also "neashe and tender to keep and somewhat skittish and fearfull, partly by nature and partly by breaking in".

Blundeville's suggestions for breeding are: great trotting horses for war;[7] ambling horses to ride; swift runners to run wagers or to "gallop the bucke" and for exercises of pleasure, or, for the countryman, horses for draught or burden.

He insists that climate is important and he thinks that the Turk, Jennet, etc., will change their natures when brought to a colder climate.[8] The Neapolitan can well prosper here, although he comes from the "garden of Italy". Then comes an illuminating remark, for, he says, these horses "do well here and stallions of these races should be mated to big mares with trotting pace for war". Use the horse of Barbary for swiftness, and he mentions carthorses both common and "exquisite", they should be good travellers, have good hooves, be reasonable feeders and preferably gelded.

Now to some advice on breeding. The breeder is "not to bestow his labour in vain" but to choose the right mare and stallion. The stallion for his beauty, his goodness and age, and to be of one colour: brown, bay, dapple gray, light bay, roan, white lypard, pure black with white star. One colour "cleare and bright", but white points make him very beautiful. He must be of goodness and health. *Never send a jade to the stallion!*

Aristotle says mares may foal all their life long, but Blundeville considered they should not foal after ten years. They should be colder of nature, very quiet and domesticated. The stallions should run with their own mares in separate enclosures where the grazing is good and with water and shelter.

Now let us see what Marcus Fugger, a German merchant and banker of the sixteenth century, has to say in his book *Von der Gestüterey* ("About Breeding"). He lived from 1529 to 1597 and was a great expert on horses. He covered all the breeds in Europe and he also had a stock farm in Hungary and in the Allgäu. The woodcuts by Jost Ammam from 1539 to 1591 in this recently republished book illustrate my description of the strains of horses at the beginning of this chapter.

It is of considerable interest, then, to learn this late Middle Ages horse breeder's opinion on the type of stallion to use to breed war horses. It is also quite apparent that *heavy* horses were unknown! He says that a medium-sized stallion of excellent proportions and temperament should be used; all others should be avoided. There are first the Persian, but not a big one. Then the Moorish (Berber) horse, but just as the Persian is big so these are small. Thus one should select the biggest of these. And third-

[7] The knight in full armour was on the way out.

[8] This is exactly what Abbas Pasha said, see page 206.

ly, there are the Spanish horses—the Neapolitan Ginetes because they are one breed.

> I like the Spanish horses best as stallions, and in Spannia they are regarded as the noblest. In principle he shall be of the same size as the mares. To believe that the stallion shall be large and the mares small—the mares large and the stallion small and that such differences in proportion can produce properly proportioned offspring is an error—although an inch or two makes no difference.

The ideal age for a stud stallion is 4½ years because he has reached mature proportions:

> Some say, he should not be used after ten years, others fifteen or twenty but . . . if one can obtain a good Persian, Moorish or Spanish stallion then take care of him (because such are not likely to be found at every Frankfurt Fair) and use him so long until one gets another or can improve on him in some other way.

Fugger advises: "If you want good foals use good mares, for seed sown in bad ground will produce fruit but it will not be so perfect as if it had been sown in good ground." Further we read that the mare should have a small, fine head, with small, pointed ears, large eyes, fine, big nostrils, small jaws, whilst the further description fits a cob! Let no one imagine that a stud of mares can be assembled in eight days. A long time and patience is needed to select the best because generally the mares that are offered are those which their present owner does not want. Fugger adds that twenty-six years earlier he bought twenty-four mares from the Netherlands which he selected on their beauty—because he knew no better—and to tell the truth only four of them were really good mares. So if one wants to collect good mares, buy two- and three-year-old maiden mares and he would hope one would be successful.[9] But there is more to a good mare than beauty of conformation.

We learn also from Marcus Fugger, as from Thomas Blundeville, of the breeds and types of horses that were bred in other countries and imported into the various kingdoms and principalities which made up Germany at that time. The beautiful woodcuts illustrate these horses and it is not necessary to let our imaginations wander and thus increase the size of the horses of 400 years ago, since every single woodcut represents a light or heavy cob of about 14·2 h.h. The only exception is an illustration of wild horses.

The first of the breeds is the Turk,

> under which name a vast number is included, for example, all those countries which the Turks have conquered. We call them all, without differentiating, Turkish horses. But there are many countries, unknown and

[9] The text of Marcus Fugger's book is delightfully written in High German, i.e. with Gothic lettering. I wish his expressions could be translated!

> untravelled, so that none of us and not even the Turks themselves can specify the horses, or even say exactly from which country they originated. Because I have seen many small, weak and poor Turk horses which are of no value whatever. But also I have seen many tall, long barrelled as well as small greys, with such extraordinary long legs and necks that nothing uglier can be imagined. But the others with broad chests and croups on short legs, with small heads and necks are the sort I like.

He adds that not many of this sort are to be found and then money will not buy them. Otherwise there is a shortage of good horses in such countries belonging half to oriental and half to northern countries such as the Karst. Many are by Turk stallions and come under Turkish breeds. For work and war they were a great deal better than ordinary Turk horses. The first are the Persian which belong to the Copts and are therefore not really Turks but they were distributed to Europe through the Turks.

These horses are excellent, of the sort not to be found anywhere else in the whole of the Orient *because most of them come from Media* (my italics!), from the plains of Nisea. And just as Herodotus says in his book, they are big, strong horses, broad of chest and croup, because, of course, the Persians wear heavy armour. The horses are seldom cut because it is thought they are more useful for battle. Not many of these horses find their way to Europe and if one has such a horse, "then he should care for him and love him. Because they are fast, active, strong and willing and possess all the qualities which a war-horse should have. Only they are untrue in the stable—they bite".

Once again, I would stress that this description of a sixteenth-century war-horse does not fit a nineteenth-century heavy horse. Later we shall see how *this* horse was developed.

Just as the Germans of those days regarded the Turk, Spanish or Neapolitan, so the Turks regarded the Persian, and if they could not buy them for money, they were obliged to use their fists!

The second to use for breeding are the Arab and Armenian[10] horses, but they are rarely seen in Europe for the Turks look on them much as they do the Persian and rarely acquire them: "These horses are not so strong as the Persian but are nobler but with bad mouths, they go well and are surer on their legs than any other Turk horse, they are kind and gentle and do not hot up like the Persian."

The third strain named by Marcus Fugger is the Moorish (Berber) which came from Africa via Italy and thus to Europe. There is a great difference between the Arab and the Berber:

> and I mention this particularly since I know well, that generally no difference is made and they are held to be the same. But if anyone looks at the countries, he will see that there are many hundreds of miles between Arabia

[10] Note the combination and refer to the text on the Arab, page 206.

and Barbary, and that Arabia lies in oriental Asia and Barbary in meridional Africa. This is where the Moorish or Berber horses come from. They are usually small, very willing and have plenty of stamina, they are very hardy, fast and can cover long distances at the canter. They are first-class horses and if only they were larger like the others mentioned there would be no better horses for war. The ancients thought a lot of these horses since they made all Africa famous. When they wanted to represent this country they showed a single horse [like Egypt and the crocodile!] the same as today we find them on old coins such as those of Hadrian. And they are the only horses that are not frightened by lions, other horses are terrified.

Furthermore, they were highly valued by the Turks. It was also said that they need no bridle but were easily guided with a stick.

As Fugger says, there were good and bad horses of these strains but generally for ordinary people the Turks were not of much use. They could not stand the cold and raw climate nor the deep roads. They were generally hard mouthed but it was difficult to say whether this was due to their long canter, i.e. they knew they had long distances to cover, or because of the weakness or strength of their backs. This point was freely discussed at this time by horsemen. Fugger's opinion is, however, that the hard mouth comes from a weak back, because as soon as a horse realizes that his back hurts, he fears a jab in the mouth but this does not cause him to obey.

Fugger then refers to the two strains of Spanish horses, firstly the noble Ginetes which he praises highly, considering them to be descended from the Moorish horses (see page 163). He also adds that these good horses are now rare in Spain (probably because they were exported all over Europe, like the later Norfolk Roadster). The other sort are the Villanos which are strong big horses and not as beautiful or as gentle as the Ginetes, and this is why they are called Villanos and not because they are used by the peasants. For work or war, they are better than the Ginetes. They are excellent for war, strong, willing, true, lively, with good mouths: "and in my opinion they beat all the other horses I know of. I know of nothing against them although other writers think otherwise, I rely on my own experience *quae rerum est magistra.*"

The best of these horses were bred in Andalusia, others in the mountains around Martos and Jaén. The horses could not be bought because twenty-five years earlier the breed had almost vanished—like the Ginetes—and were subsequently bastardized by recrossing with Berbers. If the King of Spain had not founded his stud at Córdoba with around 100 good mares then there would not be a single good horse left in Spain: "For with my own eyes I saw a nobleman pay three hundred ducats for a dapple grey Tennish [Danish] work horse, which could only be ridden under a rug and for which we would not have paid sixty thalers. This shows that all the other good horses have vanished." And I suspect that our friend means too that the knowledge of horses had got so low that the average Spaniard did not know a good horse from a bad one!

The Italian horses come in order of 'goodness' after the Spanish. They were many-coloured but they were unbeatable as good horses. There was always a war of some sort in one or another part of Italy, as in France which had a bad lot of horses and always had to find horses from outside. This observation of Fugger's is of especial interest because I have had the greatest difficulty in tracing any sort of evolution or breeding of horses in France. It also explains the need for Italy to import Einsiedeln horses (see page 178) and the reason why the Dukes of Mantua and Mailand were so happy to receive the 'Horses of Our Lady'. Horses were also trained after the Italian method.[11]

At the time of Henry I of France there were many kinds of foreign horses in France and enormous sums of money had been spent in getting them. Fugger adds that the internal wars serve no one and that the nobility were all warriors and were therefore very much occupied with horsemastership.

As far as the horses in Germany are concerned, no specified breeds are mentioned although there were a number in the Lowlands (Netherlands). Lombardy and Romania as well as Hungary supplied horses, whilst most princes and noblemen had their own studs. They were good horses "*da due felle* as one says, that is, middling horses which could be likened to the Neapolitan and were just as good". The best, however, were those brought in from Italy which did very well in German studs. The Neapolitan was not only valued and sought after by foreign countries but was highly thought of in Italy too, as better than other breeds.

He says that they were chiefly bred in mountainous or hilly regions and that such horses are infinitely better than those bred on richer soils. Be that as it may, a great number of Neapolitan horses enjoyed fame but "truthfully there were a good number of very indifferent horses amongst them and if one did not take enormous care and work on them as one did, then you can believe me their fame would not be worth much". Apparently there were two strains of Neapolitans. The Coursers were big, high horses bred in the king's stud originally because they were not much used elsewhere on account of their size and were at this time still only to be found in the king's old stud. They were used generally in carts, but before firearms became generally used these horses "did great deeds".

In Neapolitan and French armies there was a (battle) order and one can scarcely imagine what a gap one or two thousand of these horses charging could make in a line of infantry. I think one can imagine after seeing a dozen London police-horses controlling a crowd of several thousand people.

They were, it seems, enormously strong and active and Fugger himself

[11] Some 300 years before Caprilli!

witnessed, to his astonishment, many horses jumping ten or twelve jumps. I imagine he means that horse and rider were executing the *croupade.* In fact, I have not seen a Lipizzaner executing more than four *croupades* one after the other, so that any horse capable of executing ten or twelve must have possessed enormously strong hocks and back. But by 1584 only great men and potentates had any use for these magnificent parade horses which they dressed in superb harness and upon which their pages were mounted. Those horses from the king's stud from Calabria had a brand on the nearside, and those horses with an offside brand came from Apulia, but it was said that the Calabrese were the best.

Now to the second type of Neapolitan, or those which one called *da due felle* or middling horses, and there were two strains of these. The first were called *Veneten del regno*, since they were of Spanish descent and resembled the Ginetes. In fact, although they were said to be stronger, they were indistinguishable from and were clearly Spanish Ginetes and were first-class horses highly valued in Italy. The other type was a strong medium horse generally found in and around Abruz and was bred by most of the nobility carrying the brand of their owner's stud. In 1569 a little book was printed with illustrations of about half these brand marks.

These Neapolitan horses made good war-horses but they needed careful training and time to reach maturity. They were hot-headed and easily lost their tempers when wounded and could as likely bolt into the enemy as back to their own lines. So they were considered to be dangerous until they reached their years of discretion when they became more tractable. Their Neapolitan riders, says Fugger, called them foals, although they might be seven years old and over. But like all late maturing horses they lived usefully to a ripe old age.

The stud of Mantuan horses was founded by Duke Francis Gonzaga who went to great expense to bring stallions from Spain, Turkey and Barbary, and the stud consisted of Coursers, Turks, Genetes and Berbers. They were not as strong as the Neapolitan horses but they were obedient and easily taught. The reason for their softness was that they were bred on rich land, whilst the Neapolitan, as I have already said, was raised in mountainous areas. However, they were less obstinate and easily bred so that Charles V of Hapsburg[12] thought far more of them and used only Mantuan horses for his personal service.

By all accounts then, the Neapolitan Coursers must have needed some handling, and they were clearly not a breed for the amateur to play around with!

The Po valley, in which both the towns of Mantua (Mantova) and Ferrara lay, was formerly held by the ancient Etruscans, whilst later in 450–250 B.C. the fertile Po valley was overrun by the Celts. Both these

[12] Charles V (1500–58) King of Spain from 1516 and Holy Roman Emperor 1519–56.

ancient peoples were horsemen of considerable fame, so it is not surprising to find that horse breeding continued to flourish down the ages.

About one hundred years before Marcus Fugger took up his quill, the Wars of the Roses happened in England, when according to Jos. Osborne the fine horses that were bred at this period had to be dispersed. Apparently they were eagerly secured by foreign royal houses: "among whom may be particularly mentioned the Duke of Ferrara, possessor of the famous Mantuan stud". (The ruling house came to an end in 1598.) These horses were preserved and when peace had been finally established "some of the best of their descendants were re-imported in the reign [1509–47] of Henry VIII". Henry was a keen judge of horse flesh and did much to improve horses in England by rehabilitating the Eltham stud and founding those at Windsor and Hampton Court.

Osborne further says: "that he bred for racing and from Eastern sires there is proof in his trainer having been styled 'Keeper of the Barra or Barbary horses'." Incidentally the Marquis of Mantua presented him with a number of mares who were sent to Hampton Court stud.

Returning to the horses bred in Germany, Fugger recounts that in former times great value was placed on the Allgaü and Swiss horses, although, he says, he does not know where the Allgaü ended and Switzerland began! The Swiss did not breed horses except those which were called 'seven Thaler' horses, having to import great numbers, and they would not have done this if they had bred them. The inhabitants of the Allgaü were in no better situation, they bred horses for work. But the district and climate of the Allgaü lent itself to breeding good hard horses, in spite of the short summers and long winters, when all the animals had to be housed and fed that much longer. We know Fugger's affection for horses bred in mountainous districts, indeed his own 'modest' stud was situated in similar surroundings.

If the horses of earlier times were so valued, then the climate must have been milder as it could well have been. Apparently there were no German breeds as such, and it is true that most if not all of the modern German breeds originated during the seventeenth and eighteenth centuries and later.

As far as the Lowlands (Netherlands) are concerned, we find that special breeds are mentioned. Generally they were called Lowlands horses—although specifically there were the Friesian, Dutch, Flemish, Westfalen and Gelder as well as the Mountain horses. The Flemish, are Fugger says, generally fine but soft horses and there are few of them. The Dutch have a similar 'complexion' to the Flemish and there are many of these—2,000 youngsters were brought out of the country at a time—but care would be needed to preserve them as the war was not likely to help their increase. The Friesian horses were common with much feather but were reckoned to be strong, useful horses.

> But personally, I consider the Mountain horses to be the best, not only in the Lowlands but in the whole of Germany [which he calls Teutchland—the land of the Teutones] and the Westfalen are nearly equal. They are fairly big with smaller heads, necks and joints than the Friesian. The Gelder horses are long-barrelled.

I assume that these Mountain horses are none other than the Ardennes, already famous in Julius Caesar's time, and this breed fathered the Westfalen, later Rhineland heavy horse. If I am right, then they were an excellent type of heavyweight cob. Today they have become even heavier, but still retain the neat head and are fast disappearing. The Rhineland has ceased as a breed altogether.

In general the horses of the Lowlands are obedient, willing, teachable, good-tempered and some have a first-class trotting gait: "if one schools one of these horses in the Italian method and does not ruin it in its youth as is generally the case, as I have said many times before, they compare very well with other horses from distant countries. Such as I have seen in Italy and indeed owned [and they cost several hundred crowns]". So many horses were spoilt, in front and behind, because they were ridden to death when young.

The Tennish (Danish) horses which were brought into Germany were small but good horses for normal daily use. The better quality big Danish horses (the Royal Fredericksborg stud was founded in 1562 with Old (original) Spanish and Italian horses) were rarely seen and Fugger appears to be uncertain as to whether they really existed or whether they were too valuable to be allowed to leave the country, which was generally the case with Spain and Naples.[13]

There were good horses, too, in other parts of Germany, always bred by princes and ruling houses. Fugger mentions Mecklenburg, Brunswick, Saxony,[14] Hessen, Livland, Pommerania. Many foreign horses were also brought in and came, for instance, from Transylvania, Hungary and Poland: Podolien, Crabachen and Zecken horses made good saddle horses. Of course, one must remember that all long journeys were made on horseback, there were few roads and it was essential to have a comfortable mount which was capable of going long distances. The Transylvanian, Zecken and Hungarian horses were strong and fast and made good war-horses, although the Hungarian were often thin and nervy and were used in watercarts of which there "are far too many and a lot of good horses are thereby ruined, that could have been used differently".

Bohemia bred a similar good strain of horses which unfortunately had

[13] Marcus Fugger, Master of Kirchberg and Weissenhorn wrote the foreword to his book in 1527, so in fact the Royal Fredericksborg stud had not yet been founded, although there probably were good horses in Denmark.

[14] The provinces helped to found the Hannoverian breed of horses.

the habit of carrying their heads too low; probably the pastures were too rich. The horses thus became soft and were better for draught than for riding.

If I have discoursed at too great a length on the breeds of horses present during the late Middle Ages, it is because I want to dispel the myth of the non-existent strains which we have been repeatedly told *did* exist. It is only by clearing away these cobwebs caused by ignorance or fertile imaginations that we can properly set the scene for the emergence of both the light carriage and saddle breeds and the heavy draught transport breeds of the nineteenth and twentieth centuries. That they did evolve from the cobs and the mountain and lowlands breeds by selective breeding and feeding shows just how versatile the horse is. It also shows us that any particular breed can rapidly disappear if it is not cared for.

6. The Spanish Horse and his Contemporaries

Before we examine the development of the most important European breeds, we must look at the evolution of the horse in Spain. There is a bronze statue of a horse of the third century B.C. found at Despeñaperos and this shows us a horse with very high action coupled with a convex profile typical of the *caballo ibérico*. These peculiar marks of conformation appear to have originated in the Iberian peninsula—or from north Africa via Spain in very early times. They are to be found in almost every breed whose ancestors can be traced to this part of Europe.

Beginning with the primitive strain, the Sorraia inhabits a region between the Rivers Sor and Raia, Portugal. This horse is semi-wild and is descended obviously from a wild forebear—although some members of the tribe have been domesticated throughout history. The colouring is similar to the Tarpan (perhaps a little more yellow dun) whom it resembles in many respects—although the profile differs in that the Tarpan has a straight or concave face, whilst the Sorraia has a straight or convex face. Both strains have a tendency to zebra markings on the legs. With this knowledge of the conformation and characteristics of the Sorraia, the evolution of the Criollo of South America becomes apparent (illustration 18).

In this evolution, however, and in the development of the Spanish horse in particular, we must include the Garranos of Minho, which S. Ruy d'Andrade places amongst all the other representatives of *Equus europaeus*. Some Garranos have a distinctly Criollo resemblance, whilst in others it is not difficult to envisage the future Andalusian or Lusitano. The famous Velásquez portrait of Don Balthazar Carlos shows him mounted upon a Garrano pony.

Thus the horses which were taken to the New World will have been of the Garranos and Sorraia strains, which subsequently evolved into the Indian pony, Mustang and various American breeds with doses of other blood. On the one hand the famous spotted Ginetes of Ridinger and on the other hand the equally famous Appaloosa of the Nez Percé Indians.

What a lot of breeds of horses owe their descent wholly or partially to Iberian ancestors. Thus to mention only four: the Neapolitan, the

64. Prince Mauritz of Orange-Nassau, mounted on a Spanish Ginete (painting by P. van Hillegaert)

65. *(below left)* Amsterdam patrician Pieter Schout on a pure-bred Andalusian horse (painting by Thomas de Keyser)

66. *(below right)* Charles II in 1660, waiting to embark on the ship that will carry him home to England. Bear in mind that he was 6 feet 2 inches tall, so this horse is not as small as it looks

Kladruber, the Irish Draught horse and the Criollo of South America. There are others, including the hackney.

We have already seen that Spain was a leading horse-breeding province under the Romans. The simple expression *Castellan* to denote a war-horse, a horse bred in Castile, gives some idea of the value of these horses; whilst the Córdoba stud is said to date to the time of Abd-ar-Rahmans (or Abdul Rahman) (756–788). But really the most important breed was the lighter saddle horse, the Ginete or Jennet, perhaps so called from the Greek *Gymnetes*, lightly armed.

In fact the famous Ginete *was* the Spanish horse, ideal for High School and to show off a gentleman's horsemanship and always portrayed with high knee and hock action. The well-known Mantua breed also had Ginete ancestors. Indeed, "many presents of fine thoroughbreds were received by Henry [VIII] from abroad, among which special mention may be made of the *mares* sent to him by the Marquis of Mantua, which were doubtless some of the first occupants of the Hampton Court paddocks".[1]

Apart from the unique semi-wild and feral Sorraia, there existed a heavier strain known as Villanos. Some experts think that an early introduction of the cobby Noriker or Noricum horse may well have provided the genes for the later common horses. *Villano* is best translated as 'rustic', thus the Villanos horse was 'country-bred'. There was too the Garranos (not entirely unlike and quite possibly distantly related to the Garron of Scotland) and the Minho, the *Equus europaeus* of Spain.

Jaques de Solleysel, who was Master of the Horse to the King, writing in 1664 in *Le Parfait Maréchal*, tells us that Spanish horses are exceptionally beautiful and the best of all the breeds to serve a clever artist as a model or to carry a monarch in front of his people. They are not so light as the Berber horses nor as heavy as the Neapolitan but have the best characteristics of both breeds. He says,

> the Genettes have a wonderful active walk, a high trot, an admirable canter and an exceptionally fast racing gallop. In general they are not very big but there are nowhere better bred horses. I have heard extraordinary tales of their courage. [This is one hundred years after Blundeville's opinion of the breed!] There have been horses that were so badly wounded that their guts were hanging out and they had lost all their blood, but with the same courage and pride with which they carried their rider into battle, so they carried him out of the foray and saved his life, then they lay down to die since the spark of life had left them sooner than their courage. The best horses are bred in Andalusia and especially in the royal Córdoba stud, and those of Cardone and Molina are highly valued.

The Villani di Spagna, or Villanos, were larger, he says, and of more ordinary conformation and they were bred in mountainous regions.

The Spanish Court was closely connected with the Court of Austria

[1] Jos. Osborne, *The Horsebreeder's Handbook*, 1889–90.

which influenced the other European Courts, including naturally Holland, Denmark, France and Britain. The schooling of High School or Dressage horses became an end in itself and every young man of consequence and courtiers were, as a matter of course, required to be able to school their own horses. Great personal discipline and patience are needed to school a horse in the perfection of this art. They spent months under the tuition of such masters as the Duke of Newcastle (1592–1676) who, during his exile in Antwerp, established a famous riding school. In 1658 he published his first book on horsemanship: *Méthode et Invention Nouvelle de Dresser les Chevaux.* This classic book called in English *A New Method and Extraordinary Invention to Dress Horses and Work Them according to Nature* was published in 1667 when he returned to England.

The Ginete stood around 15·2 h.h. and much of his beauty for the Spaniard lay in his convex face. This characteristic was greatly admired three centuries ago, and since Spanish stallions were in demand at most of the courts of Europe, the characteristic became fairly widespread and appeared subsequently in later evolving breeds. Much of the goodness of the Spanish horse was inherited by the Dutch Hard-draver—Europe's first racing trotter. As breeders of the Hard-draver, the Dutch people became the founders of the sport of trotting, a sport that was to wing its way across the Atlantic and, with the help of a Thoroughbred stallion Mambrino who sired Messenger, the founder of American Standardbred Trotters, was to capture the hearts of millions of yet unborn Americans.

The reader's attention is drawn to page 166 regarding the very dominant gene which nicked so excellently with those of the native mares of the Netherlands. The breeding of the Hard-draver goes back to the sixteenth century when William of Orange freed Holland from her Spanish overlords. The first trotting races took place on the roads and streets and both methods of riding and training were extremely primitive.

The best known of the Dutch breeds is the black or very dark brown Friesian horse, an ancient strain that was very nearly lost. The Gelderland, too, has Spanish characteristics, and this has become a useful saddle/harness horse having evolved from the cavalry/coach horse which resembled in some ways both the French Norman and the Norfolk Roadster. Both the Norman and Gelderland have, way back, Norfolk blood—and from a century or so earlier, I suspect that the Norfolk had an introduction of Spanish blood! The Norfolk and Lincolnshire cobs which were trotters *par excellence* almost always had plain, straight faces, sometimes with an inclination towards the convex, which they could not have inherited from any *native* source. They are the ancestors of the modern hackney whose activity and trotting action are famed.

There really are so many grounds for thinking that trotting ability and action came via the Berber through the Spanish Ginete and thus to the

67. Smetanka, the founder sire of the Orloff trotter

68. Trotting on the ice. Jet-black Friesian stallions competing for the 'Golden Whip and Silver-Plated Harness' on the ice track at Laandam, Holland, January 1827. The trophies are exhibited hanging from a pole. From a painting by W. Verschuur

breeds whose characteristic gait is the trot rather than the canter. Some writers believe the trot to be an artificial gait, the walk and the canter and perhaps the amble, being the natural method of progress of the species *Equus*. But there is nothing 'artificial' about a stallion high-tailing round a paddock when he wishes to impress his mares—his action is almost invariably a high exaggerated trot!

The Berber horse has a number of trotting records to his credit. In 1886, a 14-year-old Berber horse harnessed to a heavy four-wheel cart with two passengers completed a journey of 109 kilometres (62 miles) in eleven hours. Ben Daoud rode a Berber mare a distance of 160 kilometres in nine hours. Lt. Carrèz rode his Berber charger 420 kilometres in three and a half days. And, as Count Wrangel says, "Compared to the Arab, the Berber can develop a very satisfactory speed at the trot."

On the trotting tracks of Algiers and Blida, genuine Berber horses achieved the following results: Sultan in harness—4,500 metres in 8 minutes 7 seconds. This equals 1 minute 50 seconds per kilometre. A horse called Mabrouk achieved 1 kilometre in 1 minute 48 2/5 seconds, and Octave 1 kilometre in 1 minute 48½ seconds. The trot is coupled with leg action or activity as opposed to the canter which is a gliding action. If a horse whose normal gait is the canter is encouraged to trot, this gait is then a gliding action, in common parlance known as 'daisy-cutting'.

There is the dominant gene which produces the convex profile of the Berber, perhaps in Mesolithic times it evolved in the Iberian peninsula, and which was transmitted to the Spanish breed, which breed produced the markedly high action required for High School Airs. This horse was esteemed throughout Europe for this action which was governed by his powerful hocks and we know that this blood later helped to develop the coach and carriage horses. Thus the same action and profile was passed on to the great roadster and hackney breeds of horses including the Irish Draught horse. A few of the cob breeds still retain traces of this ancestry whilst the hackney is valueless without the exaggerated action, which, of course, is brought out by training. Through the Irish Draught horse, the Cleveland Bay too, and the countless hunter dams with a bit of coach blood, our national and international show jumpers, trace back to the Spanish/Neapolitan breeds, as also do the Hannoverian, the Holstein, Anglo-Norman, Maremma, and many others.

One famous private stud of this period, Murau in the Steyermark, was owned by Prince Adam Franz of Schwarzenberg who twice held office of Chief Stud Director to the Emperor. Today he would probably hold the office of Master of the Horse. Prince Schwarzenberg, like a number of his contemporaries, bred for colour. At Murau, he bred the much favoured Spotted horse, which was so especially treasured a little later by the Nez Percé Indians of the Palouse River, in the New World. Within the Austro-Hungarian Empire (1510–1867) there were already the

Royal Carst (later Karst) Stud, the Imperial Halbthurn Stud in Hungary, in which Charles VI took an especial interest since he had not inherited it but was himself responsible for founding it, the Imperial Stud of Bohemia at Kladrub, and the Stud at Lipizza. The first and the two last-named studs were to give their names to breeds of horses, one of which, the Lipizzaner, was to carry the name into the twentieth century. At first there were little or only slight differences in the horses bred at these studs—unless some were bred specifically for the imperial stables for coach work as coach and carriage horses. Eventually use determined the type of horse to be bred. Lipizza had to supply carriage and High School horses, Kladrub heavy coach horses, but this imperial stud of Bohemia produced at least one entire who was an "*extraordinari*" jumper! Halbthurn Stud bred royal parade horses and possessed a number of original Spanish horses. Amongst these was His Imperial Majesty Charles VI's favourite saddle horse Hermoso, bought in 1717, which was later retired to stud there. Another favourite of the Emperor's was the Spanish stallion Sincero. The only original Berber stallion to be used at Halbthurn was the chestnut horse Brilliant who had three white stockings, a white coronet and a blaze. Apparently he was a genuine Barb with slightly convex face, from Tripolie, and not an Arab passing for a Barb![2] Some of these stallions were painted by J. G. von Hamilton and the portraits hang in the Rössel room at Schönbrunn. They were fairly large animals. One such was the dark dun Andalusian who stood 176·4 Stm (17·1½ h.h.) and was at stud at Halbthurn for many years.

Joya, a 'golden brown', was also an original Spanish Riding School horse; he was sent to Halbthurn in 1719, where he became the most successful sire there. The cream grey Philosopho from the Imperial Carst Stud was equally successful at Halbthurn from 1725 onwards. Another, the black horse Moro, interestingly enough a Spanish horse from Sardinia, was also much used; although apparently these stallions were at stud for only four months of the year and they were sent to the studs where their services were needed, from Vienna or Laxenburg. All horses carried the brand of the stud where they were foaled, and some, in addition, the royal C or J with crown (Carl or Joseph). The latter brand was in use before 1711.

Cerebo, a skewbald or odd-coloured horse from the imperial stud of Bohemia, was a famed jumper 'springer'. He stood at Halbthurn from 1718–23 and produced a son of the same name. Another stallion at stud as an original Berber of scarcely more than 14 h.h. with a desert brand mark. He was grey and had the dark legs, mane and tail associated with the Bornu breed, from the district south of Lake Chad in present-day Sudan. Hamilton also painted a Spotted horse (in 1704) and this was one

[2] It is about this period that the name Berber was substituted for Barb in the English language.

69. A rouncy, depicted by Albrecht Dürer

70. *(below left)* Albrecht Dürer's knight, fifteenth century. The horse shows the influence of the Spanish horse on the cob strain

71. *(below right)* A palfrey (from a painting by Breviario Grimani)

of the earliest horses of this colour. He belonged to the Schwarzenberg Stud at Murau. It was very fashionable to breed for colour, and grey, dun, palomino and odd colours such as piebald and skewbald were very popular.

The Fredericksborg Stud in Denmark was one of the most important in Europe and it supplied grey and also white or cream horses for royal ceremonial occasions. The Fredericksborg Stud played a part in helping to found the Hannoverian Cream coach horses—the forerunners of the present Windsor Greys, which are still used for all royal ceremonial occasions in Great Britain. The Windsor Greys are not a breed and there is no blood relationship between the two famous coach horse types—the Hannoverian Creams and the Windsor Greys. There were several teams or pairs of the famous Creams at the Court of Saxony. Apparently the King of Denmark, Frederick IV, stayed at the court of August the Strong at Dresden on his return from a visit to Italy and he probably presented the King with horses from his own stud. This is confirmed by another painting of a white or cream horse, held by a groom carrying a pair of pistols which belonged only to the more opulent and ornamental tack of the Fredericksborg Stud. The horse carries no brand mark on his neck, which would have been the case if he had been bred in one of the imperial studs.

The stud books date from 1718. Horses were sent direct from the Spanish stud to Karst or elsewhere as they were needed. But unfortunately the stud books do not give any information as to the pedigrees of the stallions so that one would assume that the breeding lines of the sires were not taken into consideration. More attention was paid to the mares, as was also the case in other European studs at this time and, indeed, also today. The Bedouin thought more of his mares than of his stallions. Only in Britain and later in the United States is the reverse the case, the mares often receiving less attention in a pedigree than do the sires.

O. Antonius suggests that blood lines and pedigrees came 'into action' roughly a century later when geldings were used as saddle and work horses and few entires were kept and then only for stud duties. Mares were kept out at grass but a teaser was used and covering was done by hand, i.e. the stallion was never turned out with his mares (a reversal of Blundeville's policy). Many mares appear to have remained barren and the loss of both mare and foal was common. Mares were not used for work. Veterinary medicine was unknown and many horses died from accidents, with abortion, pneumonia and other lung diseases which could destroy a complete stud, and infections and fevers including tetanus. Gassebner says that planned breeding for horses for the army began under Maria Theresa in 1763, and the army studs of Waschkouz, Mezöhegyes and Babolna-Kisbér were founded.

In the stud book of Halbthurn in Hungary, which was situated within the Austro-Hungarian Empire, we read of the number of mares as at 18 May 1733. The entries are as follows:

> Names of the mares, from whence they came; dams and sires; colour marks; shape of head; age; height. By which stallion covered in 1732; foals born; by which stallion they were covered in 1733; how many foals each mare has had; which of these was retained or discarded; how many at present at the stud; how many colts from each mare were later sent to Vienna; how often the mares were barren; what foals they then bore.

It is presumed that this detailed record was kept for the Emperor personally. There could well have been more details; no mention is made of the stallions except for their names. There is also no account of the day-to-day running of the stud. However, we learn that there were 428 brood mares at Halbthurn for the length of the stud's duration, 1717–43.

From other papers dated 1733, dealing with His Imperial Majesty's Karst Stud, we read of mares of the Liechtenstein breed, and these include four black mares. In 1756 many papers were lost through fire so that the pedigrees of horses bred later can only be traced back to 1770. On one page there are entered fifteen stallions of 'Courser type' and twelve stallions of riding or saddle type, of which the former produced fifty-two foals and the latter thirty-eight foals, whilst from the four jack asses six mules were born. None of the stallions appears to have been overworked whilst at stud, although a further note shows the Director's anger that one of the riders[3] had ridden a stallion to death.

The Duke of Lothringen also owned a private stud and frequently exchanged horses with those of the Emperor, especially when his stud was removed to Austria in 1736 and was united with Kladrub in 1836. The Lotheringen Stud Koptschan on the River March was created by Charles VI's son-in-law, Franz Stephan, later Emperor Franz I. There is a list of those horses which His Royal Highness transferred, including twenty-eight colts and fillies of three and four years old, the dams and sires of same, their type and colour and so on.

The Halbthurn Stud lay to the north of the mountain country of Austria and was called at the time by its official Hungarian name of Feltorony. The country is flat and stretches to the Neusiedler Lake and beyond. The latter is a genuine salty steppe lake and much of the surrounding land is similar to steppe. Antonius describes it as a marvellous country for bird life, and the climate is considerably more Continental than in Vienna 80 kilometres distant. This would mean that the summer is much hotter and the winter consequently colder.

An earlier castle was destroyed by the Turks in 1603. In 1711, Count Alois Harrach built the present castle, the architect being the famous Lukas von Hildebrandt who also built the Belvedere Palace in Vienna.

[3] *Bereiter*, a designation still in use today for riders of the Spanish Riding School in Vienna.

The Emperor Charles VI became the owner of the estate and stud in 1717, and lived among his horses. He was always present during the autumn when the mares were culled (rejected as unsuitable for breeding), and the young colts, which he intended for the Spanish Riding School, were paraded in front of him. Charles loved this district, especially for hunting; he was the keenest of all the Hapsburgs. With his death in 1740, Halbthurn Stud came to an end.

Halbthurn was founded in 1717 with thirty mares from the Harrach Stud, forty-seven brood mares and ten foals of varying ages from the Kladrub Stud in Bohemia, and nine brood mares from the Karst Stud at Lipizza. Amongst the thirty mares only eighteen had been foaled at Halbthurn (then Harrach), others had come from the Count's Bohemian studs and from a stud in Liechtenstein. Two mares were bought from the postmaster in Raab, although they were not retained for long. Some of the Harrach mares were Neapolitan Coursers—and these were subsequently sent to Bohemia to breed mules! Kladrub was the only imperial stud where mules were bred in apparently fairly large numbers. Two to four jack asses were kept.

> Big strong mules were required, not only for pack purposes but also for the highly useful sedan chair and they were as equally indispensable as the necessary number of harness and saddle horses. The Emperor Charles IV was responsible for giving mule breeding a wider basis when he caused 20 big Holstein mares to be bought for Kladrub. (Antonius).

Antonius writes that there was yet another reason why the Courser mares were not retained at Halbthurn. Breeding of harness and carriage horses was just as much outside the plan as was the breeding of mules. The colour of some of the mares is interesting; most were black, some brown. Two Liechtenstein-bred mares Misca and Mia (in 1712) were 'mouse-grey' or blue dun; another mare was also dun. *Myszha* in Polish means 'little mouse'.[4] Either they were dissatisfied with the colour or the foals were not up to standard since Misca, Mia and Medusa were cast."

> Another group of Halbthurn mares requires notice, because of their breed designation. The chestnut mare Fatuna by Visier × Ancora was called a Turk. So that we may assume that the sire brought Turk blood into the breed . . . Perhaps his real name was Vezier not Visier and he was captured during Prince Eugen's Turkish siege. Many of his descendants were chestnut. (Antonius.)

The original Halbthurn mares were not especially successful and quite a number were eventually weeded out. The chief line of the new stud consisted of mares of Kladrub origin. Twenty-one were given as Spanish, fourteen as Neapolitan and one as 'Turk type'. The latter was the pale

[4] As stated elsewhere, the Tarpan is also blue dun or mouse dun. It has a coat exactly the colour of a mouse.

dun mare Volpina by Real (Regal?) out of Quabia. This mare was one of the most successful brood mares at Halbthurn. In 1719 and 1721 she also foaled two dun odd-coloured colts to the skewbald stallion Cerebero. Later she foaled four times to Curioso, two to Montebello and once to Superbo—and at the end she was sold to a stud official for 3 florins. Her daughter Inamorata was also a successful brood mare.

As we know, the famous Byerley Turk—to whom most Thoroughbred horses trace back—was also captured at the Relief of Vienna in 1683. As a result of the Turkish invasions decades later, a considerable number of captured Turk horses (of what strain we do not know) found their way into Europe via Vienna. Even in 1598 after the victory at Raab as many as 700 well-bred horses were captured. The Emperor received fifty of the best. Thus it is not surprising that some Eastern blood found its way into the imperial studs of Austria. What is surprising, as Antonius says, is that there was so little of it!

Kladrub horses are shown as of both Spanish and Neapolitan breeds, the latter as 'saddle type and jumper'. There does not appear to have been a basic difference between the Spanish and Neapolitan riding horses, only the Neapolitan 'Coursers' or 'Runners' were larger and heavier. "The division of Kladrub saddle horses into Spanish and Neapolitan types is not a sign of different breeds, but simply the direct descent from original stallions of this or that origin." (Antonius) In Halbthurn there appears not to have been a division of horses into harness or saddle types, there were a considerable number of Karst mares but not a single Kladrub, so it seems that Halbthurn was intended to breed only riding horses and that it had relieved Kladrub of this duty. Later Kladrub was to supply not only the grey coach horses for the royal mews but the famous black heavy horses, whose last official duty was to draw the coffin of the Emperor Franz Joseph to its resting place in 1916.

The Karst Stud was, of course, the home of the Lipizzaner saddle horses. In the stud book there is no special mention of type, but perhaps they were an already established breed. Generosa was one of the best of the brood mares. There were a number of others, such as Tacha, Dalmatia, Palma, Perosa, etc. The colours are important. There were thirty-one black mares, thirty-one brown or bay, and Fatma the Turk was the only chestnut, though she was joined by her chestnut half-sister, thirteen greys, eight duns and two odd-coloured. The much-favoured exotic colours are absent, for there are no Spotted horses, nor Creams, only two piebald or skewbald. This is a contradiction of Gassebner's text, where it is stated that during the fifteenth century the imported Spanish ancestors of the horses at the court studs were responsible for breeding the later much-favoured cream-coloured horses. The Lipizzaner stud book also supports the theory that exotic colours were not looked on with favour at the Austrian Court. Even Creams were not liked, although later they were bred, since the colour was so popular in many private studs. Subsequently 123 brood mares are entered, amongst them 23 blacks, 66 brown or bay, 2 chestnut, 4 dun and 25 grey; amongst the foals there were, as might be expected, all colours.

Kladrub had many dark horses: blacks, black greys, browns and only five chestnuts and duns.

Lipizza supplied mares at different times to improve the breed in Halbthurn and sometimes mares from private studs were brought in. Amongst them were six brown mares from Prince Schwarzenberg's Stud at Murau—one of the largest and most important in Austria. Prince Adam Franz von Schwarzenberg was, as already mentioned, Chief Stud Director to the Emperor Charles, by whom he was accidentally shot and killed during a hunting party.

The stud of the Elector of Mainz also supplied some mares, whilst the princely stud of Liechtenstein also supplied some black mares and four of these were of varying heights from 13·2 h.h. to 18·1½ h.h.!

In 1733, when the Court had come to regard exotic colours rather more favourably, the 'Collubrad' stud (Count Kolowat?) supplied three dun skewbald mares and one light dun. The first Spotted mare appeared in 1734—a red roan called Fontura II. Possibly she came from the Karst Stud, since Hamilton's painting of 1724 (which hangs in the study of the Director of the Spanish Riding School, Vienna) of the herd of mares at Lipizza shows a Spotted mare. A number of piebald mares were also added from the Plettenberg Stud (from north Germany?) and in 1743 thirty-nine mares from Count Tättenbach's Stud 'Wisl in Steyermark'. Antonius says that Count Josef Ferdinand von Remstein and Tättenbach was the last of his line and that his estate lay at Raun, now in Yugoslavia. From the number of mares it would appear that he had disposed of his stud. As we know, the Emperor had died in 1740, and shortly after the addition of these mares the Halbthurn Stud itself came to an end.

However, Lipizza continued and flourishes up to this day. It seems that a number of the Steirmark (Steyermark) mares were 'cream but nearly white'—and two months after their arrival in Halbthurn, thirty-eight were specially selected and sent to the Karst Stud.

The war with Frederick the Great caused the Kladrub Stud to be evacuated to the Island of Schütt.

One of the best and most valuable stallions, Joya, must be mentioned; he produced sixty-three foals. He was a golden brown Spanish horse with white hairs. Another was Philosopho who sired at least thirty foals; he was a Spanish-type grey horse from the Karst Stud.

Another well-known stud of this time was that of Lippe-Bückeburg not far from Hannover. This stud supplied Nobile de Lippa (1728) whose descendants played a considerable role in the breeding of later Lipizzaners. Nobile founded the Lipp-Nobile line.

If we study the pictures in the Rössel room, it seems fairly certain that Hamilton portrayed the different strains of horses as they really looked. At this time the Roman nose was regarded as desirable and, indeed, as beautiful. The exaggerated concave face of the Arabian horse had not then become popular and many horses, both mares and stallions, had straight faces. Nagler writes that Hamilton understood the characteristics of the different breeds as did few of his time. It was fashionable to clip the ears of horses in order to make them look smaller and in some cases it is not easy to discern if the ears of Hamilton's horses were *painted* to look smaller, or if some had been deliberately shortened.

The conformation of School horses of the period was strong, strength and muscle being needed to enable the animal to execute the difficult and strenuous school movements, such as the Airs above and on the ground—courbette, ballotade, levade and pesade.

Just as the Lipizzaner today is strongly made with broad chest, broad and strong croup, and deep and long body, so were the horses of Spanish, Neapolitan and Fredericksborg strains well and strongly made.

To sum up, all these horses possessed similar *characteristics* of conformation. Today we may like to imagine that the Neapolitan (from the stud near Naples) with the pronounced convex profile was a special breed. This is not so; the breed had the same characteristics of conformation as the Karst or Kladrub, unless a few of the horses foaled in these studs had Berber blood, when obviously they would be smaller and lighter. The stud in particular gave its name to the strain but the strains were all interrelated, including the Fredericksborg and other horses bred in the top private European studs of 250 years ago—and the oldest and best was the Spanish horse.

Some breeders preferred the one type—the Neapolitan or Kladrub—others the elegant High School or the 'Campagne' type, which was bred with a cross of Berber or Turk blood, or in a few cases with stallions from 'Engelland'. There was no particular breeding according to line or pedigree, as today with the Lipizzaner according to ability for a particular Air, and in the Thoroughbred according to blood lines and families and racing achievements; although colour was selectively bred as with the Fredericksborg and the white/creams of Hannover.

Whereas we associate all Lipizzaner horses as being white, or rather grey—the foals are all born very dark or black—(because those generally used and preferred for display or exhibition at the Spanish Riding School and on tour recently in the United States and Great Britain are grey) in fact the Lipizzaner may be any colour. Indeed, these horses are not only used and trained for High School but they make excellent harness horses and in Yugoslavia they are also used on the land.

We have already seen (Chapter 2) that white or grey horses were greatly favoured by monarchs, ruling princes and powerful lords from

Asia to Europe. The influence of white—purity and light—became a remarkable equine symbol for the House of Guelph, the ruling princes of Hannover. The Electress Sophie bred and kept at her summer residence, Herrenhausen, a special strain of Cream horses. This very small and specialized breed was started in 1730–40 when the cream entire Auguste V and the silver-grey 'le Barbe Blanc', together with the Danish stallion 'le Blanc', stood at stud. It was several years before the white or cream colour became dominant and it was even more difficult to breed stud material. Colour breeding to result in the desired colour *and* a sound useful hard horse is a most difficult undertaking. Unhappily, from the Hannoverian end, this experiment was doomed to failure. Auguste V was born dark grey and became ice-grey. His sire was the Barbe Blanc, dam la Deèsse was light grey by the little Vienna Turk, bred in England. Auguste V could be used at stud for only three years and was then so ill with fistulous withers that he had to be put down—with a dagger, then the only method of destroying a horse. It is said that the Danish horse le Blanc was born white. He went to stud in 1753 but of his fifteen descendants only one is mentioned. Le Royal Danois and some mares were also added to the breeding stock, whilst Hampton Court supplied 'la mère Angloise' and her daughter of unknown pedigree.

Thus the Cream Hannoverian horses had two blood lines: Danish (Fredericksborg) and Berber (Barb). They were small horses with many faults, records the chronicler, not the least being infertility (if it really was a fault and not partly the result of bad management i.e. lack of vitamins and minerals in the grazing and feeding).

By 1790, in-breeding had produced no better results. But before that, the Master of the Horse informed the king in London: "There are only 6 or 7 cream mares in Hoya and we expect only one cream foal. We cannot allow this characteristic coat colour to disappear and beg, therefore, to be permitted to buy 6 cream mares. Kuhlmann, the horse dealer offers such cream mares for 180 Reichtaler per head." The king replied: "Your list to hand, you have permission to buy 2 mares." A mare called Charming was sent from England. But the number of pure-bred white or cream mares was low and generally they were albinos, grey roan or red roan. At the Great Lodge, Windsor, Stud in 1803, there were eleven stallions and seven mares. Some of these had come from Mecklenburg where they had been evacuated (presumably from Hoya or Herrenhausen because of Napoleon's invasions). Later the number rose to 59 horses from Hannover and 111 horses from Memsen and Neuhaus in Hannover. Not all of them were cream.

In 1815, a number of mares were sent back to Hannover and the Prince Regent became interested in supporting the continuation of the breed. In 1870 the stud director in Berlin informed the King: "Even with in-breeding it is ever more difficult to breed the Cream horses. We have great difficulty in getting the mares in foal and then the foals are small, weak or crippled and it is rare that we get a sound foal.[5] If the breed is to continue a few grey mares and stallions must be bought in." When Finance Minister Scholz refused to buy two grey Arabs from Austria, the end of the Cream horses in Hannover was in sight and the last of the mares was shot soon after foaling a cream foal which died shortly after birth.

In England, however, the position was a great deal happier. I suspect that the grazing was better, not to mention the stud management. The Crown Equerry, Col. G. A. Maude, C.B., told the Prince of Wales: "Since Hannover became independent of England, we breed our own Creams in England. We have never imported stallions and we practise in-breeding but strangely the horses are growing larger. We have four brood mares."

There is a delightful story told of Napoleon's victorious march into Hannover. The horses and the silver had been evacuated to England. Even so, in 1803 Napoleon managed to acquire a Cream Hannoverian horse and took him to Paris. He intended to ride him in the victory parade—but the stallion refused to let him mount. Napoleon persevered and the horse then bolted with him.

The most important sire in the creation of the Cream breed was Adonis who caused a furore at the coronation of Franz II, when ridden by the ambassador from Hannover. 1,000 taler was offered, a huge sum for those days, but he was not for sale and in 1793 was sent to England, where he became the favourite of George III.

Another sire of importance was Ido: "When the royal visitors arrived the stables and horses took priority to be visited. Maj. Schweppe rode Ido whom he had trained to stand on his hind legs and walk the length of the riding school—in this manner he represented the coat-of-arms of Hannover."

We have little idea today of the extent of those European studs of 200 years ago. By our modern standards they were *vraiement formidable.* Marbach under the King of Württemberg owned a modest 1,000 hectares (a hectare measures nearly 2½ acres); Lipizza 2,500 hectares; Trakehnen 6,000 hectares; Mezöhegyes 17,000 hectares; and Radautz covered a distance of 120 kilometres (around 75 miles) and contained twelve studs and five farms.

[5] This seems to show a truly remarkable lack of knowledge of animal management, even for those days, and even though the science of genetics has advanced enormously. Albinos do have a weaker constitution and thus in-breeding could prove fatal. Duerst maintained that the white or cream Fredericksborg horses carried a lethal factor which, he thought, may have been coupled with the white factor. In fact, Pulos says that the dominant white gene is lethal in the homozygous condition and causes an early abortion. Thus the breeding of the Creams was genetically unsound.

In his excellent book *European Studs* Frederich Traut describes Mezöhegyes, founded in 1785 by the Empress Maria Theresa with military requirements in mind. From the original 262 brood mares, 22 stallions and 90 provincial stallions, within twenty-five years the number had grown to 4,142 brood mares spread over nineteen studs, of which a great proportion were *wilde* studs. Wild in this sense can be taken to mean feral. Altogether, he says, there were 13,300 horses, which were reduced to around 1,855, although even at the end of the century 500 young entires were contained in one rearing stud. (When I was last in Poland, I visited one such stud where 200 entires were being reared.) A huge number of cattle were also raised at Mezöhegyes and there were farming industries of one sort and another which in all employed 7,000 people and contained between 18 and 20,000 animals.

The interesting thing about this huge complex where space was of no importance (compare the vast, confined, cattle-rearing factories which are projects of twentieth-century agriculture) is that it is not possible to keep together permanently too large a number of horses. The reason is that with such huge numbers, proper management was impossible, illness became rife, and especially illnesses of an infectious nature. Nature is not a factory and there are limits—to exceed these limits is to invite disaster. This experiment was made three centuries ago. With all the devices and advances in veterinary science, equine influenza is *still* rife!

Many studs were founded and owned by cloisters and monasteries, for instance Kladrub or Einsiedeln, founded in the year 1062 which is testified by documents. The monks were generally recruited from the nobility and landed gentry (as were the knights) since there were few occupations open to younger sons, but they brought horses with them from their parental estates. The horses then belonged to the cloister. With gifts from princes and great landlords, unworked land and poor soil were brought into cultivation, all with the help of the horse.

Long before Blundeville was born, the Schwytzer had become well-known horse breeders. Thus in 1310 they drove a herd of horses on to chapter property and fed them on hay stored there!

In 1964 Cloister Einsiedeln celebrated a thousand years of horse breeding and then (as now) there stood in the Cloister stables many fine horses. They were admired at home and abroad. The Abbot Gerold describes a journey:

> With a magnificent train, the Abbot ventured south in 1464 to Italy. The reason being a visit to Pope Pius II which had taken weeks and months to prepare. The dramatic incident in Firenzuola was unexpected by the Abbot and his following. As they entered the town, young and old crowded to see them. They had never seen such well turned-out riders nor such beautiful horses. The squires, fearing that the admiration was too excitable to be genuine, prevented their admirers from entering the quarters appointed to them.

This action turned into a proper scrimmage. Called to their help, the city fathers managed to break up the tumult. With divided opinions the visitors from Einsiedeln left Firenzuola for the south. They regarded their mounted escorts with affection.[6]

Breeding of Einsiedeln horses reached its peak between 1500 and 1800. Indeed, as we have seen, this was a period when many breeds of horses reached the summit of excellence and usefulness. People had time to sit and think and work out what they hoped to achieve. The abbots of monasteries—and there were a great number right across Europe and England—were men of learning and science and they influenced the surrounding districts. It is our spiritual and material loss that these centres of learning no longer exist. For it was in these places that progressive steps in agriculture and horse-breeding took place. Mendel himself (1822–84) was Abbot in the Augustinian monastery of Brünn. Abbot Conrad III was responsible for increasing the numbers of cloister horses, and the Dukes of Mailand and Mantua showed such an interest that many splendid horses found their way over the Gotthard Pass to the sunny south. Their new owners called them "Cavalli della Madonna" (the Horses of Our Lady).

In 1655 the types of horses consisted of saddle horses, cart horses, pack horses, brood mares and stallions. When the stables were rebuilt in the eighteenth century they were said to have no equal as buildings of architectural impressiveness. The motto was "breed for stamina". This is still the aim and it was achieved throughout the ages with the help of fresh blood from Italy, Friesland, Spain and Turkey, although always the genuine Schwytzer blood was preferred. The Einsiedeln horses stand about 15·2 h.h. and will go in harness, work on the farm, carry one through forests and over the mountains or show jump internationally. This is the story of one cloister stud—there were many more.

So we come now to France which had evolved some old strains prior to dilutions of Berber, Spanish and the blood of the horses of Attila (Hun, and therefore partly Mongolian, etc.) and the horses of the Moors. How much or how little French breeds of horses were really affected by the latter invasions is difficult to judge.

Within the former Duchy of Normandy (1154) there were the cantons of Mortague, Mondoubleau and Courtalin, the cradle of the old breed of horses and later an area which became known as *Le Perche aux bons chevaux*. The present Percherons resemble not at all their ancestors, who were of less massive conformation and generally brown in colour. In fact, the early horse of Le Perche looked much like his neighbours, the Breton and Norman cobs or Postier, as they were called. Several centuries later both Brittany and Boulonnais provided basic material for breeding

[6] *1,000 Years of Horse Breeding in Cloister Einsiedeln* by P. Albert Huber (EDE Verlag, 1966).

larger horses and it is from the latter that the grey colour is inherited and became predominant, although there are black Percherons, particularly in the United States.

French experts, amongst whom was Eugène Gayot, refer to the improvement in the old breed which was due to Oriental blood. In the year 732, 300,000 Saracens mounted on Arab and Barb stallions under Abdul Rahman invaded France. They were met by Karl Martell between Poitiers and Tours and overthrown, and many first-class horses were taken as booty and eventually crossed with native mares, so the foundations were laid, if not the breeds individually, for a number of new strains which were to make France a leading horse-breeding country in the centuries to come.

Karl Martell's cavalry were armed knights and this is the earliest date, as far as I know, where the *dextrarius* appears, and with ultimate success, for the lighter Saracen cavalry were put to flight. Later on when the Crusaders returned home bringing with them Arabian stallions, a further improvement in breeding took place, also with additions of Andalusian blood.

In time, in France as well as other European countries, there was a great demand for draught horses and Le Perche breeders found it necessary to increase weight in their light active breed of horses. To do this they brought in the heavy Breton and Boulonnais and so for a time the Percheron breed appears to have lost its beauty of conformation and activeness and to have become coarse, plump, slow and bad-tempered. Those responsible for the management of Percheron breeding tried in vain to prevent the old posting cob from dying out, and they had partly succeeded when Americans appeared on the scene who were prepared to pay enormous sums for the new colossus, and this ensured victory for the new breed. Two thousand stallions and mares were exported annually between 1879 and 1889.

Edmond Gast, writing in 1889, says: "This so useful, so necessary light draught horse which provided excellent service to the heavy parade equipages, omnibuses and the French artillery, no longer exists."

Luckily the export trade died out and with the increasing interest in the sport of trotting and therefore in breeding horses that could trot, once more the old type of active Percheron was sought after, and Wrangel writes that there are records of Percherons gaining a kilometre record of 1·58 seconds, and equally able to trot 3·5 minutes to the kilometre over 10 to 15 kilometres pulling a heavy cart.

In 1897 the American market was once more actively interested in the now regenerated active Percheron which proved to be an excellent cross for the better-class native American mares. Nearly one hundred years later the black Percheron is still a great favourite in the U.S.

From the earliest times Normandy was the greatest province for breeding horses in Western Europe and both warm- and cold-blood breeds thrived on the good pastures. Norman overlords and wealthy cloisters saw to it that the horse population was cared for. From the beginning of the Crusades "quality Arabian" (*sic*) stallions accompanied the returning knights and one of them, Oliver de Rohan, brought nine home with him. Also, it was usual to donate one's war horse to a cloister which would care for him and probably use him to help improve the local strain which was usually of the cob or pony type.

According to Wrangel, breeders, whether nobility or clerical, bred for endurance. In the present day and age the word has almost lost its meaning. But in 1238 near Saint-Lö a public race-course was built and until the reign of Louis XIII (1610–48) the breeding of the Norman horse improved. The wealth of the nobility was, of course, a factor and most men of consequence kept from 20 to 150 horses; as, indeed, did many European nobles on their estates, up to the Second World War!

Queen Elizabeth I sent a company of Scottish Foot soldiers to Henry IV in return for forty horses of the Norman breed. Unfortunately Cardinal Richelieu brought to an end any further progress of the State Studs and a number of valuable sires found refuge on private estates, until the requisitioning armies of Napoleon Bonaparte arrived on the scene when every Norman horse, mare or filly was taken. In 1806 the State Studs were reorganized and on Napoleon's orders Egyptian, Turkish and Mecklenburg stallions were bought so that disorder in breeding was added to disorder!

It is worth while quoting again from Wrangel. Before the revolution Normandy raised four completely different types of horses and the situation was very similar in some of the other provinces and departments.

1. In the districts of Courtonier, Seez, and Alençon the Merlerault breed of horses was to be found and these were descended from the old Norman-Armorican strain which was descended from Eastern horses brought in by the Moors, and replenished with further Eastern blood when the Crusaders returned.
2. The Contentin breed which was made up of a black and a brown family. The former were considered to be top class cavalry horses and were bred in the Arrondissements Valognes and Saint-Lö. Legend said that the ancestors of these horses were Danish-bred, having been brought in by the Northmen when they conquered Normandy. The brown family was descended from crosses between Danish [presumably Fredericksborg], Neapolitan and German stallions mated with native-bred Norman mares. They were bred around Vire and not much good can be said of them.
3. Around Valoges and Cherbourg a small active fast breed of horses was to be found called la Hague, whose ancestors had been brought from Sicily when southern Italy was conquered by the Normans.
4. The cart horses: *bidets d'allure* of Bessin, which were famous for their extraordinary hardiness and active pacing gait.

So there existed in France before the Revolution the basis for breeding and raising good horses. But the breeders themselves appear to have been missing and many opportunities were lost. With the founding of the Jockey Club in 1833, the trotting races which began seven years later and the arrival of Napoleon III on the throne, horse breeding improved and has never looked back.

Both strains of Norman horses—the old Postier cob and the Norman saddle horse—have been regenerated and in 1972 were seen in London. Some of their regeneration was due to the blood of Norfolk Roadster stallions which the cob resembles, whilst the Norfolk Roadster, one of the most valuable of British breeds, was allowed to become extinct. France's Trotter breed owes a lot, too, to a Norfolk Roadster, Norfolk Phenomenon, imported 1851. A number of imported Thoroughbred sires were used to improve the native Norman breeds, the Trotter breed, as well as laying the foundations so that France was to become one of the leading countries of the world to breed the Thoroughbred horse, and perhaps to a lesser degree the Anglo-Arab. Clearly throughout France's history many strains of horses have come and gone. Amongst them the Limousin and the Nivernais, which could trace its origin to the draught cob of the old Morven breed which was greatly esteemed for hunting in the seventeenth century. During the last century the Nivernais was an all-round saddle and harness horse.

During the years 445–53 the Mongolian hordes under Attila the Hun overran Europe from Denmark to Hungary, from the Caspian Sea across the Rhine. It is not possible to say whether this invasion influenced Western European breeds of horses. Some Eastern European strains, especially in Hungary and the Balkans, got an admixture of Mongolian and Tatar pony blood.

One of the oldest strains is that of the Camargue cow-pony whose antecedents are said to go back to prehistoric times. One cannot help feeling, however, that this horse has certain characteristics in common with both the Sorraia of Spain and the Berber. But it has had little, if any, influence upon other breeds of horses.

During the second millennium A.D. in Europe more and more attention was paid to breeding horses for use in war—where the Asiatics left off the Europeans took over. The war horse of the armoured knight appeared, then the courser and carossier, the cavalry and artillery horse. Horses were bred of varying degrees of swiftness or up to certain weights, from those of the scouting units, dragoons, hussars, guards, troopers and police horses. Horses were wanted in ever-increasing numbers for sport: hunting, polo, racing on the flat and over fences, show jumping. Spahis in Tunisia were mounted on 14 h.h. Berber horses; Eastern European scouting regiments had equally small extremely hardy horses, and the Indian regiments had small horses of Arabian type. Horses that were required for the various royal stables were of the heavier coach calibre, such as the black and grey Kladrubers, the coaching Holstein and Hanoverian including the special cream horses of the Electors of Hannover—whose work has been taken over by the Windsor Greys (which are not a breed). Some countries like East Prussia and Ireland produced first-class all-round saddle horses. The former[7] of native, eastern and Thoroughbred descent, the latter from native pony, Spanish, Irish Draught and Thoroughbred ancestors. Other countries like Poland, France, Spain, Austria and Hungary, produced elegant strains with eastern ancestors, Arabians, Anglo-Arabians, Shagya, Gidran Nonius, Carthusian, Andalusian, and Lipizzaner.

[7] See *Flight of the East Prussian Horses* by Daphne Machin Goodall (David and Charles, 1973).

7. Horses for Transport

In Great Britain, towards the end of the seventeenth century and the beginning of the eighteenth, the turn of events encouraged the creation of the heavy working horse. The invention and use of firearms in warfare caused the vulnerable *dextrarius* to become obsolete and rapidly contributed to the reorganization of lighter, faster cavalry in the armies of the European states.

Then there was the necessity to transport increasingly larger and heavier quantities of goods and merchandise. Pack horses were now no longer adequate to carry the wares which preceded the period of the Industrial Revolution. At first the highways were tracks made up of mud and stone and teams of horses had to be used in waggons and in the heavy wooden coaches too. The increased use of roads led to John McAdam inventing the surfacing of roads with granite. In 1816, he successfully 'macadamized' the roads around Bristol.

During the centuries preceding, it was always the smaller horse that was used on the land, if oxen were not used, and the heavier animal was ridden, but now the light farm horse was not up to the demands to move weight and the phrase "weight to move weight" came into effect. The demand for a heavy dray horse became ever greater. The famous horse fairs were those of Malton, Ripon, Lenton, Rowel, Harborrow, Melton, Northampton and Pankridge. The Duke of Newcastle wrote:

> I am very ignorant of the West Country where my Lord Paulett's Ancestors had a good Breed of Horses; and by Chance now and then my Lord of Pembroke did Breed, but I never heard of any Rare Horse of his Race. In Worcester and in the Vale of Esam, there is good Strong Cart-Horses; in Cornwall there is good Naggs,[1] and in Wales excellent good ones; but in Scotland the Galloways are the best Naggs of them all.

Dr Fuller[2] refers to the horses of Hertfordshire: "their Teams (oft times deservedly advanced from the Cart to the Coach) are kept in excellent equipage, much alike in colour and stature, fat, fair, such is their care in dressing and great feeding of them". A servant was brought before a Justice of the Peace accused of stealing grain. The man brought

[1] 'Nagg' refers to all-round ride/drive/dray horses.

[2] *History of the Worthies of England*, 1662.

along his team tailed together, since, he said, if he were the thief, they were the receivers.

The horses of Denmark, Holland and Germany are represented by the Duke of Newcastle as specially adapted for coach, cart or plough. And Holinshed says:

> Our cart or plough horses (for we use them indifferently) are commonlie so strong that five or six of them (at the most) will draw 3 thousand weight of the greatest tale with ease for a long journie . . . Such as are kept for burden, will carie foure hundredweight commonlie without any hurt or hindrance . . . Of such outlandish horses as are dailie brought over onto us I speake not, as the genet of Spain, the courser of Naples, the hobbie of Ireland, the Flemish roile,[3] the Scotish nag. (*Chronicles*, 1577)

Since the *heavy* horse really came into his own during the nineteenth century, I refer to the *Cyclopaedia of Agriculture*[4] which specifically states that the cart horse is not an original breed in Britain but was imported. It says: "Recourse was had to the large black horse, which had been known throughout the fertile plains of Europe from time immemorial . . . during the reign of the Edwards, repeated importations of these animals took place." Since the prevailing colour was black, a distinct strain came to be recognized in England under the appellation of the Old Black English Cart Horse which was distributed throughout England and was divided into three kinds: the "heavy, massive animal, reared in the rich marshes and plains" of the Midlands expressly for London brewers; the smaller but still fairly heavy horses used in agriculture, strong compact and slow; and a lighter, more active animal which possessed either an admixture of blood of a lighter breed, or was the descendant of the Flanders coach horse, and being discarded from the carriage to make way for the Cleveland Bay was of more use at the plough-tail.

A further interesting observation on the effect of the mineral content of the soil on animal production, the *Cyclopaedia* insists that the dray horse can be reared only on the richest pastures and "is found in the fens of Lincolnshire in the greatest perfection". Colts were bought at two years old and worked moderately until four being well fed during this time, *indeed previous to their re-sale they are often taken out of work and fattened in loose boxes*, almost like oxen for slaughter, and in this state they were supposed to please the London brewers. A note of warning is added that this redundancy of fat can cause various diseases.

"These ponderous animals are frequently 17 h.h." Some Shire horses reached 18 h.h. Obviously these massive, majestic animals walking purposefully through the London streets would cause passers-by to pause and look at the turn-out and at the brewery signs on the dray!

They often suffered from the ills of bad feet, side-bones and weak-

[3] Roile: mud; a horse capable of going through mud.

[4] *A Cyclopaedia of Agriculture*, by J. C. Morton, Vol. II, 1855.

72. *(above left)* Sixteenth-century cobs, ploughing and harrowing

73. *(above right)* A pack rouncy or cob returning from a stag hunt; early sixteenth-century Flemish (both from calendars)

74. *(left)* A sixteenth-century cob from a stained-glass window

75. Ladies driving a pair of ponies (medallion from the German Cloister Fischbeck Tapestry, 1559). The ponies were probably descendants of Tarpan stock

76. Only light horses or ponies were used for draught in the Middle Ages. (From the Duc de Berry's Book of Hours)

77. Ancestors of the Norfolk Roadster and the Hackney, and particularly of the Suffolk Punch. These working cobs appear in Nathaniel Kent's *Agriculture of Norfolk*, 1794

78. An illustration of pack ponies from *Yorkshire Costumes*, 1813

nesses generally. And this is where the old saying probably originated that a good horse had "neither cart nor black blood". Not all were black, however. Some were bay, brown, grey and even roan. A few were odd-coloured—and here possibly Villanos blood may have been responsible (and who knows, perhaps even the Ghora Spotted horse had contributed to *their* ancestry!).

So the horses were bred principally in the Midlands and clearly it was here that the Shire breed of heavy horses evolved.

At this time, 1855, the Suffolk Punch could be sorrel and bay as well as the more general chestnut. Their origin lies deep in their native county, but references are found to the "old breed" and I am sure that this referred to those cobby all-round horses which were used by the farmers of both Norfolk and Suffolk. Sometimes there cropped up an animal such as Blake's Farmer, a famous Suffolk Punch sire foaled in 1760. He appears to have been chestnut in colour, not much above cob size, and he was a trotting horse. No doubt he stamped his conformation and his activity upon his foals.

They are, the *Cyclopaedia* adds, exceeding staunch to the collar, free from hair on the legs and have nice heads, their height ranges from 15 to 16 h.h. These chestnut horses are still used on some East Anglian farms and on these heavy clay lands; it is very necessary that they are clean-legged. Heavy horses with much feather, as a preponderance of hair on the legs was called a hundred years ago, often suffered dreadfully from grease (sores on the legs and heels).

As for the Clydesdale, it was greatly valued in the one-horse carts used in the valley of the Clyde. Provided they were well fed, they were active, willing workers, though sometimes 'hot', and perhaps this defect could be attributed to the mixed parentage of Dutch ancestors. An Earl of Huntingdon imported some Dutch mares and since there was no *breed* of this appellation, the mares could have been of mixed ancestry. These mares were crossed with the native breed which was probably the old pack-horse strain and in this way the much-valued Clydesdale heavy horse was created.

Perhaps it might be interesting to examine the method of feeding these massive working horses. One farmer fed each horse six pounds of oats at five in the morning, and at noon the same again, and at 7.30 p.m. forty-seven pounds of steamed food which could be potatoes or turnips or a mixture of both. In addition, each horse consumed every twenty-four hours fourteen to eighteen pounds of fodder hay or straw. All horses were fed chaff as well. On principle, horses that were regularly worked should not be turned out to grass; although, as a child, I distinctly remember that the working horses were turned out every Sunday.

M. A. Denham, in *North of England Folklore* (1858), tells us that the last Bell horses were kept by Charles Michell of Forcett Hall, near Rich-

79, 80. Also from *Yorkshire Costumes*, 1813: Yorkshire pony peat-gathering; and East Riding or Wolds waggon drawn by local horses. This was before any improvements to the breed took place

81. England's Glory, a brown horse bred in Lincolnshire, 1814—17 h.h., weight 180 stones

82, 83. *(below)* Suffolk Punches, nineteenth and twentieth centuries

mond, Yorkshire. The bells were suspended on a wooden framework, which was covered with a parti-coloured worsted fringe. The horses' trappings were handsome, and the bell was suspended at the neck of the leading horse of the pack trains, by which the Salters used to carry their merchandise over the moors.[5]

Horses of all descriptions, types and breeds (there were not so many breeds then as now!) were sold at fairs and markets up and down the country, and Long Melford in Suffolk had not only one of the most famous but also one of the *longest* horse fairs in the country. Horses were tied on both sides the length of the village street which is over a mile long.

During the course of the following decades, British heavy horses were exported to America and Canada and other countries which required heavy dray horses to further their expanding economy. Percheron horses were imported from France whilst Belgium continued to improve her heavy Brabant breed. The Brabant was responsible for various heavy German breeds such as the Rhineland, Silesian, etc., but now many heavy European breeds have disappeared, just as the heavy black horse disappeared many decades ago.

Comparatively the life and usefulness of the heavy dray or working horses has been very short. They were selectively created for a purpose, and once that purpose no longer existed due to the fact that mechanism took over, the day of the heavy horse was done. But dray horses *are* still used by brewers, even in the 1970s, on the streets of London, and for short journeys they are much cheaper to work than diesel or petrol-burning lorries. Shire horses *are* still bred in the Fen country and, I am told, their numbers are again increasing with exports to the United States. The Suffolk Punch also still works on a few East Anglian farms.

Like some of the light breeds, discussed in the next chapter, perhaps these magnificent equine giants have become a status symbol or an advertising symbol for large concerns who can still find a use for dray horses. Certainly the appeal to the spectator is great, for nothing catches the eye more than a team of brown or grey Shires or chestnut Suffolk Punches trotting into a show arena.

[5] This is where the expression 'bell mare' originated—a mare who controls and leads any bunch or drove of horses. And in Switzerland the 'bell cow' still brings the herd down from the mountains at milking time.

8. Light Breeds, Thoroughbreds and Trotters

Having established in the earlier chapters of this book that the wild horse is a most conservative creature and of choice inhabits a limited territory, we come face to face with one of his most baffling characteristics, namely his evolution into a domestic animal and his versatility under domestication. This versatility under domestication in the hands of man is surely such an antithesis, a contradiction of a deeply ingrained characteristic, his conservative shyness and natural wildness, that one is at loss to explain it. I do not think that anyone has ever seriously or scientifically tried to.

Of all the prehistoric animals[1] without exception—or almost without exception—the horse (and here we must exclude his cousins the zebra, onager and ass, all of whom have retained their primitive colouring, conformation and naturally wild spirit, the domestic donkey excepted, and have been domesticated only in a more restricted form) is still with us in his primitive prehistoric form, e.g. the Asiatic wild horse. His descendants and the descendants of earlier wild horses have flourished under conditions of extremes laid upon them by man's dominion.

The Asiatic wild horse grazes on the saxaul and salt grasses of the steppe, just as he did thousands of years ago, but the Arabian horse in the tent of his Bedouin master lives on dates; the Shetland pony on the barren seashore of northern islands eats fish and seaweed; the brood mare contents herself on highly developed grasses, whilst the highly trained international show-jumper lives on a diet of concentrates in the form of small hard nuts or cubes. Change the environmental conditions and any one of the horses mentioned above will survive happily on the diet of the other.

Take any horse or pony out of his normal environment and routine and send him 5,000 miles by air north, south, east or west. Not only will he survive but once he has had a few days to rest and to acclimatize him-

[1] There are, of course, several surviving prehistoric species, including the elephant, an equal in domesticity but not in versatility, and others which have not evolved appreciably, such as the alligator. But none has been domesticated under such varied conditions, forcing a complete change and reversal of its natural pattern of life.

self, he will win an international race or jumping competition *against* horses which have been bred and trained within a few miles of that particular venue.

If we keep a horse on the road travelling from show to show from week to week, he will still try to jump a clear round over painted artificial obstacles. He will try in spite of an abscess and toothache, in spite of fatigue, hunger or sore shins, and in spite of the fact that his rider is sometimes the biggest hindrance of all.

It is surely this versatility that saved *equus* from extinction. Many times during the course of prehistory and history, extinction must have been very near indeed. At no time or period could there have been large populations of horses as compared with man. In early historic times and at the present time small wild minorities lived and still live and reproduce in isolation. Even today, when the human inhabitants of so many countries are so pro-horse that they breed them for every conceivable purpose, there are estimated to be only about 65 million horses and ponies throughout the world.

During the past 300 years *equus* has evolved by selective breeding into many breeds and strains and it is my intention here to examine the most important of these. These are the horses which may affect future evolution and for this reason, the major pony breeds, important as they may have become in the twentieth century, have not been specifically included. We have seen how they all evolved and it is to be hoped that fervent breeders will not change a type by introducing outside blood—but the ponies themselves have had little effect upon the *history of breeding*!

Count Wrangel[2] states that the best of all the Eastern Russian breeds is the Teké-Turkmene, and Sir S. Baker, in his book *Clouds in the East*, says that these horses look like genuine Thoroughbreds and often stand about 16·1½ h.h. They are always kept out in the open, covered in blankets and hoods which rub off the mane.

W. T. Lyell, who spent a lifetime in Asia, wrote in *The Field* that the expression 'pony' covered all Asiatic breeds including the Arab, with the exception of the Turkmene which averaged 15–16 h.h.—an almost maneless horse built like a racehorse and capable of covering tremendous distances. This stamina was developed by the thieving habits of the inhabitants, who were of necessity often obliged to cross waterless steppes over distances requiring many days and this without any other food than that carried by the rider.

The horses seem to have been able to gallop 100 miles for six to eight days on end and were fed only on balls of milled barley, rye and sheeps' tail fat! At other times, completely covered by rugs, they were fed barley

[2] *Die Rassen des Pferdes*, 1908.

84. Shire horses

85. *(below left)* A Clydesdale horse from Scotland

86. *(below right)* Percheron stallion, a French breed, 1972

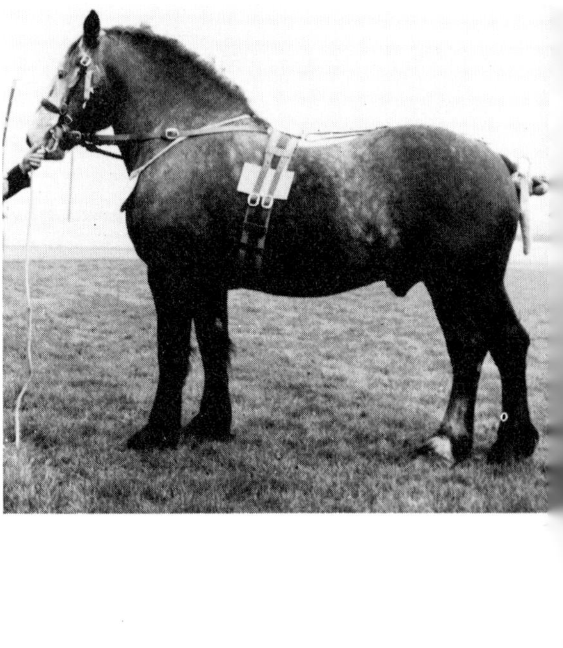

and chaff, and in spring time, like the Persian and Turkish horses, they were "freshened up" with green corn or grass. They have, says Wrangel, big plain heads, often ewe necks, and the ears are unique, small, pointed and turned inwards, which characteristic the Turkmene horses appear to have passed on to Indian breeds. We know already that the Sogdian horses of Alexander the Great were famous and that in particular Marco Polo noticed them during his journeyings.

Of course, new blood has been introduced as when Timur (1336–1405), whose power reached from Mongolia to Egypt, gave the Turkmen people 5,000 mares, and later Nasr-Eddin presented 600 mares. These were presumably of Arabian stock. "Because," writes Wrangel,

> in spite of this, the present-day Tekke horse shows no characteristics of the Arab; it looks more like an English Thoroughbred; it is big, thin [herring gutted!] with small joints and chest; its neck is long and thin, with a marked wither, heavier head and the hindquarters not particularly developed. The particular characteristic of the Arab—the high tail carriage—is completely absent. The Tekke horse often has a sloping croup and thus the tail is badly set on. The face is concave or at least straight and heavy; on the other hand, the eye is remarkably large. Besides being good gallopers, they have a three tact gait.

The Altaier (Kalmuck inhabitants of the Altai Mountains), originally the Telenget, together with the Teleuts, are typically of Turkic origin and have a very ancient language, living amongst the Mongolian people. He is a *passionierter* horseman and very awkward on foot (*schleppend und wackelend*), in fact, he wobbles as he drags his feet if forced to walk anywhere! But the moment he mounts a horse, he becomes a free man: "his eye clears, he sits easily and it seems as if new blood flows through his veins. Horse and rider are one". (Wrangel) These people owned, at least up to fifty years ago, a first-class type of horse, generally light in colour and often completely white. Herds consisted of 20–60 animals of 8–25 mares, 5–10 geldings, 5–15 yearling foals, 2 two-year-old horses, a few three-year-olds, and a stallion.

When the writer Radloff visited the Altai about a hundred years ago, many owners possessed numbers of such herds and a certain Altaier, Toloj, owned as many as 6,000 animals. The mares are milked at evening. So that horses are recognizable, some families have special marks and horses are branded on their quarters. Colts are gelded when two years old with the knife which the Altaier carries with him. Geldings are then separated and used for riding at four years. Horses were sacrificed in spring and autumn and had to be light-coloured. Neighbouring Jakuts live on horsemeat and kumiss.

Schwarz tells us that the inhabitants of Turkestan possessed horses of the Argamake, Karabair and Kirghiz breeds and a strain of cross-bred milking mares. The few Argamake and Karabair were owned by the wealthy, all others were used as harness, pack or saddle horses by the ordinary folk.

The horses were fed on barley, dzungaria and lucerne. The latter was the main food for most horses and it gave them large bellies and generally did not help conformation. Lucerne has been grown since time immemorial and was always the main feeding crop for horses and cattle. It grows on all oases for at least ten years without re-seeding. The Kirghiz themselves are descended from the Huns and also, I suspect, from the ancient Saka. The Kirghiz-Kaisaken kept thousands of horses for milking since the national drink is kumyss. Foal's meat is especially liked.

"The noblest and hardiest of all Turkestan's horses is the Turkmene, which is the equal of the Arab in beauty but exceeds it and every other breed in the world for speed, stamina and hardiness." Schwarz owned a Turkmene horse for ten years and says that for size and height its equal is not found amongst other breeds. The Turkmene horse[3] has long thin legs, long body, long thin neck, small head, narrow chest and a generally small body and it consists chiefly of skin, bone and muscle. Even with careful feeding it never puts on fat, has almost no mane and the lack of mane hair has nothing to do with covering the horses with thick woollen blankets. This horse is exceedingly intelligent and has a marvellous memory for locality.

Schwarz's horse was once stolen and thieves always take such horses far away so that they are not caught. After fourteen days the horse returned, with a bit of chain round its neck, covered in dirt, dust and blood and starving. It had made its way back to Tashkent through all the thousands of back streets until it reached home. It also knew the homes of all his acquaintances and if the coachman was not quick enough to prevent it, would turn in at any one gate and upset the carriage. It was also extremely dependent upon its owner. During raiding expeditions, robbers would sometimes put 1,000 kilometres, across waterless deserts in five days, behind them.

The Tekke Turks usually had horse races through trackless desert between the towns of Kysyl-Arivat and Heok-Tepe between sunrise and sunset—the same distance as between Munich and Stuttgart, 220 kilometres. A race of 130 miles! Probably the Turkmene would not beat a European sprinter in speed, but over distances there is no horse to equal them, since they reach their maximum speed after galloping 10–20 kilometres (8–16 miles) and have got warmed up. They are first-class jumpers and take all obstacles with ease. The Turkmene horse in question jumped a six-foot garden wall as a six-month-old foal. Schwarz adds that they are undoubtedly descended from the Asiatic horses of the Persians, which Herodotus mentioned in Alexander's campaign. The Asiatic plain,

[3] See *Horses of the World*, by Daphne Machin Goodall.

87. A modern Turkmene horse, noted for stamina

88. Arab stallion from Egypt standing at Marbach, Germany, 1972. The Munighi type, intermediate between Turkmene and Thoroughbred

89, 90. Andalusian horses bred by the late Señor Ruy d'Andrade (whose information on the Spanish horse has been of immense value to the author)

where these horses were bred, probably around Merv, is the present-day Turkmen steppes. Much later Turkmene horses were crossed with Arabs. And Tamerlan and Nadir-Shah left behind thousands of Arab horses.

Through hundreds of years of training in waterless deserts the Turkmene horse has become what it is today, a hard, fast horse of great stamina. This is also due to the way of life of the people: Massagetae, Parthians, Sakens and the modern Turkmen people who were forced to live by raiding, because of the poverty and infertility of their land. These raiding expeditions demanded from horse and man the greatest effort and stamina in order to survive.

The Turkmen never sell their horses and only a few taken in conquest at Chiva reached Russian Turkestan. Again Schwarz says, the Argamake horses are the best and are most prized amongst the wealthy but are of little use except as show horses. They are very fast over short distances and although beautiful in conformation are probably a degeneration (*sic*) of the outcross of Turkmene and outside blood (Arab?).

The Kirghiz are the commonest-looking but have similar stamina and are closely related to the Russian Cossack horse. They are small, strong, low-built with rough coats and rough, thick manes. Their conformation is not improved as they carry their heads and necks horizontally straight forward—not like the upright head carriage of the Turkmene and Argamake horses. This head carriage is clearly the result of always having to hunt for food on the sparsely covered steppe when the head is always carried in a low position. Thus a habit has caused a change in conformation. But it is just as hardy, hard and fast as the other breeds. Never with a roof over its head, it bears the burning heat of the Turkmen summer and the 40° Centigrade cold of the Siberian winter. It lives on whatever it can scrape from below the foot-high snow and often in snowstorms and ice it comes through with no fodder at all. In winter when springs and rivers are frozen over, it 'drinks' snow and in summer in great heat and drought it can go three or four days without water. The horse can cover 120 kilometres daily without difficulty although it has only steppe grass which it has to find during the night for itself, since it never sees oats or barley.

The Karabair breed has evolved through crossing the Turkmene and Kirghiz. They are generally of medium conformation, have pretty heads, strong necks, broad chests, round barrel and generally are well made. The settled population of Turkestan value these horses for their good looks since their way of life does not require much of their horses.

In the mountains of Turkestan there is a local breed of horses owned by the Kara-Kirghiz, probably evolved from the Kirghiz strain. They are small with very compact conformation, have strong, hard feet and are quite remarkably tireless at considerable heights where ordinary horses would scarcely move because of the thin air, and by their expertise and sure-footedness at mountain climbing.

The Kashgar horses, of which there are few, are really only ponies and differ very little from those in Europe. The little-known Kalmuck horses of Russian Turkestan are a first-class breed too. They are found in the Ili valley and in other districts where nomadic Kalmucks live. They are smaller than the Kirghiz horses and are kept in a completely wild state and when wanted are caught and used for riding. They have an easy, comfortable canter which they can keep up for two hours without a break. Schwarz used them on his 1880 expedition through Kuldja and Dsungaria but adds that they were extremely fearful and this was dangerous because they got into great panic at the sight of donkeys, camels or carts. When he rode into Kuldja, so great was the panic that everything, stalls, stallkeepers, and shoppers, were scattered right and left. One horse that was used for pack transport had to be tied up like a pig before the saddle could be put on. Kalmuck horses have a compact, round body and short, strong legs. As far as their ability to withstand hunger, thirst, heat and cold is concerned, they probably exceed the other breeds. The usual gaits are the walk and canter, but many horses amble giving the rider a most comfortable ride over long distances. The Russians call this gait the *chada* or *trapata*.

The Kirghiz-Kaisaken also occasionally use donkeys, especially the caravan leaders, because they make do with exactly the same food as the camels. Their economy is based on their sheep. A Government survey carried out in 1850 produced the following figures:

Horses: 1,600,000	Cattle: 1,160,000
Camels: 390,000	Sheep: 11,350,000

The Turkmene horse has an especially sociable nature: "It surprised me to see often thousands of horses in charge of boys of 12–15 years wandering to new grazing grounds." I note here that high up in the Chilean Andes I saw a herd of ten horses in charge of a small boy of about nine years old, many miles from the nearest *estancia* or even gaucho hut. But Schwarz continues: "If a horse through sheer exuberance should break away then one of the herdsmen mounted on a fast horse with the help of pole and sling would soon catch it." Behind the sheep, horses and cattle, all in separate herds, come the camels with the entire wealth of their owners packed on their backs.

In this way, during the course of thousands of years nomad horse breeders of the central Asiatic steppes have lived unhindered by convention or politics, but with a long tradition behind them. During these many centuries, the strain of horses has changed very little. The type is similar to that illustrated on page 197.

Sultan Baber in his memoirs recounts that Kasim-Chan who was

Chan of a single Kaisaken horde[4] was in a position to put 300,000 soldiers in the field. Since these men will have been horsed, the mind boggles at the millions of horses that must have been available to the potentates of the vast Turkestan steppes—and the tremendous breeding programme in the shape of thousands upon thousands of *tabuns*.

The Turkmen people live in the land that was once occupied by the Massageten, Parthians and Sacae from whom, with other admixtures, they are descended.

Modern Mongolian strains are under 12·2 h.h. and a certain Chinese chronicle mentions that the smallest horses are found in the east, i.e. Korea. In the Pelstiligulf there was a strain with the Chinese name 'Horses under Fruit'—small ponies under 10 h.h. (100 Stm.) which could easily go under trees laden with fruit.

Ponies are reasonably cheap in Mongolia and large herds are still kept. A man's wealth is counted by the animals he owns. Racing is almost a business. Prizes are a sheep or goat or just for the sport. Races are endurance tests of five to ten miles, sometimes twenty miles. The ponies are ridden by boys who beat their animals from start to finish.

R. C. Andersson describes a journey:

> There is no water high up in the mountains, so it was necessary to carry enough in bags to last for 24 hours. During that time the ponies would not be watered at all. They were tied up at night so that they could not graze, as the Mongols said it would make them thirsty to eat the dry grass. About noon the next day we would descend to some stream or spring. Then the ponies would be allowed to drink their fill and graze for an hour or more. The Mongols trained their ponies to go without water for 24 hours and seldom let them drink oftener than that even when it was possible to do so.
>
> No true Mongol pony is over 14 h.h., and those of the desert, where grazing was poor, were seldom up to 13 h.h. Yet they carried their riders up and down mountain slopes so steep that a man could scarcely walk; they were as sure-footed as goats and at the end of a day of gruelling climbing in the heat did not appear to be overtired. The Mongol pony represents a distinct type, although variations occur in different districts, largely dependent upon feed and other conditions.
>
> It is a general opinion that he is a direct descendant of Przevalsky's wild horse—differences between them being due to the introduction of other blood. Certain ponies show an undoubted Arab strain, probably being a survival of the Arab horses brought to Mongolia during the time of Genghis Khan.
>
> They are never fed grain in Mongolia. They graze the year round even in winter, finding just enough vegetation to keep them alive.
>
> In China they feed grain and adapt to change of climate, altitude, food and general living conditions in a very short time. They make excellent polo

[4] 'Horde' actually means a given number of persons:
Great Horde 1,000,000; Medium Horde 450,000; Small Horde 800,000; Inner Horde 150,000.

and race ponies, and in winter are used for cross-country hunting and steeplechasing. Their ability to carry a heavy man over a point-to-point course of 8 or 10 miles with 50 to 60 fences, compels the admiration of everyone who has seen them perform.

Chinese ponies have not changed much during the last fifty years. The strains of ponies are similar from province to province—some being more common than others—and most appear to be ill-used. There are no native breeds of horses now comparable with those of the T'ang Dynasty.

Andersson continues:

> The terrible blizzards of the Mongolia winter take a yearly toll of about 20 per cent of the ponies but the stronger survive and increase the endurance of the breed. The herds seek shelter in ravines and press close together for warmth. The hair reaches a length of five or six inches,[5] and then the ponies look almost like bears. The herds are often attacked by wolves, and then the stallions gather the mares and foals close together, and galloping round them fight the wolves viciously with teeth and hooves.
>
> The head is large, the forehand heavy proportionately to the quarters and the chest deep and the neck and legs short. It is not a beautiful animal but "handsome is as handsome does". There is little or no attempt to improve the breed of ponies, stallions running with the herd and serving the mares indiscriminately.

Possibly the modern breeder of the world's finest horse—the Thoroughbred—would not appreciate the appositeness of the description. Yet seventy years ago it certainly fitted many Thoroughbred horses and "handsome is as handsome does". Both the Teké and the Thoroughbred have more records for speed and endurance over race-courses to their credit than any other ancient or living breed of horses; the Teké cover enormous distances too. And there were three strains: the Tekkes (Achal Teké), Goklands and Yamuds which were particularly fast. The modern Achal Teké is well known, as also the Jomud (Yamud), but I do not know if the Gokland is still bred.

The Kirghiz treat their horses much the same as the Kalmucks although circumstances are different and the horses of the former would appear to be of better quality. These are nomad Turkish-Tartar peoples, the Karak-Kirghiz and the Kosak-Kirghiz, and the latter are to be found in hordes in the steppe from the lower Volga to the Altai and from Syr-Sarja to Tobol and Irtysch. The climate varies from minus 38° centigrade to a summer heat of 63°, and the day-old foal endures both in the open, parted from its mother throughout the day and allowed with her at night only after she has been milked. It is a climate, says Wrangel, in

[5] In the museum at Punta Arenas, Chile, I found a horse skin with 6–8 inch curly hair, bright chestnut in colour. The horse had been found—the only one of its species, alone in the mountains of Tierra del Fuego, forty years earlier! No one knew how it got there or how long it had been alone.

which one can lose both sight and hearing. The horses are about 14 h.h., but their unbelievable ability to cope with the terrible cold and heat, hunger and thirst, has produced a constitution of iron coupled with speed and stamina. The colours are particularly worth mention since they are usually palomino, flecked brown, roan, very light chestnut and odd-coloured.

The Kirghiz maintain that a foal tied to a post, unable to survive or cope with the hot sun, lack of food and tormenting flies, is not in a condition to survive the steppe winter! And to amuse themselves they run races of anything from 15 to 45 miles.

At the beginning of the century, the Kirghiz suffered financial setbacks which impoverished their breeding of horses. Eventually the Russian Government stepped in and now, in the later decades of the twentieth century, both breeding and management have improved.

Near Orenburg an English firm set up a breeding establishment around 1900 to improve steppe breeding horses and to obtain kumyss and fermented mare's milk. This establishment was managed by a Dr Carrick. The mares were milked five times a day and produced a gallon of milk per day each. The breeding of mares for milking dates to around 1000 years B.C. In the eighth century B.C. Hesiod mentions this practice as applying to Scythian tribes from the steppes of southern Russia. Not every breed of horses is suitable, since the mare must be able to supply her own foal too. Fresh mares' milk, like asses' milk, is very similar to human milk and therefore more easily digestible than is cow's milk. It is excellent for delicate babies and is regarded as a cure for whooping cough in Finland. Kumyss is specially prepared from mare's milk and is used to cure tuberculosis and stomach illnesses. In Burjät Mongolia, there are no less than seventeen special farms for the production of kumyss. The mares used at this present time for milking are Kirghiz, New Kirghiz, Baschkir, Kasakh, Trotters, and the Orloff-Kirghiz cross-bred. The foal should be between forty and sixty days old before the mare is milked, and U. Neuhaus states, too, that the mare is milked five times during the day whilst the foal sucks at night.

In Siberia, the Mongolian breeds, i.e. the Amur strain, have an even more terrible winter to cope with. The foals are born out in the open far from human care and have to endure the Orkame snowstorms which rage across the Transbaikal steppes in March and April killing and crippling many animals. Then they are plagued by starving wolves although a good herd stallion will keep most at bay. "On the stallion's warning, the mares turn with their heads towards an inner circle, protecting the foals whilst the stallion gallops round striking at every wolf he can get near, and thus killing it with well-directed blows."

Horses in Siberia suffer even more from virulent anthrax which causes ulcers from which many die, especially those that get it first. Later, some are saved. When Wrangel wrote, there were still, in spite of this, a vast number of horses in Siberia. In 1950 the number given was around 13 million. Then, good racehorses and trotters fetched a high price, not surprisingly as von Hutten-Czapski states that Siberian Kosacks are accustomed to cover a distance of 700 kilometres, from Akazyn to Zuruchai in four days. It has not been possible to find out if such journeys are still covered by horses. Probably not, because there is said to be a useful air service throughout Siberia.

Nikolaus Gogol tells us a little of the history of the legendary as well as the often abused Cossack riders. During the fifteenth century, in this half-wild corner of Europe, when ruling princes had deserted them after being overrun by Mongol hordes, nothing was left to the inhabitants but the courage of despair. Everything else had gone—homes, families, lives. Thus out of the ruins and facing constant danger, a peace-loving Slav developed into a cruel and often undisciplined Cossack, and a thoroughly thoughtless horseman with no appreciation of the use of cavalry or tactics. He was not a marvellous rider in any sense but where others might endure deprivation, the Cossack never understood the meaning and was thus invaluable on given occasions—but his misuse of his horse finally brought an end, according to Wrangel, to the horses themselves. The Cossacks would seem also to have descended from the nomad Saki (Ko sak) hordes of earlier centuries and to have possessed as much or as little knowledge of horses and breeding as other primitive peoples.

With more or less knowledge on the part of owners and breeders of the requirements of selective breeding, since except in the wild studs few horses were mated or foals produced without the prior consent of their owners, breeds of horses began to take on some sort of co-ordinated appearance and characteristics of conformation. The more civilized the country whose people had a traditional inherited knowledge of equines and equestrianism, the better the breed that was created. The courts of Europe from the sixteenth century onwards produced such horse masters. Some worked unknown and unheard of on their small private estates, others became state stud directors with considerable influence.

Two hundred years ago, European Russia had the most primitive conditions of any country and the extremities and hardships suffered by the millions of horses are best described by Count von Hutten-Czapski.[6] These horses are described as nondescript nags but had one characteristic, inherited undoubtedly from their nomadic steppe forebears, and this was the ability to endure both cold and hunger and continue to exist. This single characteristic was responsible for their not becoming extinct but evolving further, arriving eventually into a decade of more enlightened keepers and breeders.

[6] *Geschichte des Pferdes.*

In the villages, herds were not kept separately according to age or sex and foals of both sexes bred indiscriminately, thus crippling their own growth. In the same way weaknesses and faults were passed on. Two-year-olds and even yearlings were put to plough and at three no work was spared them. There was no form of even the most primitive or basic stable management. On returning from work in the fields, neither dirt nor sweat was removed; the wretched animal found his only relief by rolling in the dusty or muddy earth or muck if no rain came to wash off the filth. The state of the horses from week to week can be imagined and their colour only guessed at. Stabling was unknown[7] and apparently the horses wandered around the village searching for the odd handful of straw to eat. Puddles provided drinking water.

The answer was always given that this was what they were used to and was the reason for their hardiness, but, of course, no one could assess their sufferings, illnesses nor the shortness of their lives. The treatment received by the Kalmuck horses is possibly even worse, and considering that the Kalmuck himself is fed, clothed and drinks kumyss—in fact, owes his whole existence to his horse—one concludes that these descendants of the Saki tribes are not traditionally horse breeders.

One of the earliest efforts to 'improve' a native strain happened to the Mesen in the province of Archangel as follows. In 1711, Peter the Great sent into exile a certain Prince Galitzyn, who had enjoyed the favours of the Czarina Sophie. The unfortunate cavalier, of course, found his existence boring in the extreme and begged permission to move his stud of Danish horses from Moscow to Archangel, and by selective breeding to improve the primitive native ponies. Galitzyn seems to have been as unlucky or as naïve in breeding horses as he was in courting the Czarina. His over-anxious 'improvements' resulted in the extinction of the last true-bred Mesen pony which resembled the Universal Finnish horse, and that was the end of Galitzyn's experiment.

This sad little story does, however, bear a message for all would-be breeders and 'improvers' of breeds from which so many strains of horses and ponies suffer today as they have in the past. The message is spelt out: to examine first the antecedents of the equines in particular, quite thoroughly over a period of time, to acquaint oneself fully with the dominant characteristics and the faults, to analyse nutrition and climate in which the strain has so far evolved. Then to use a cross with similar genes, not to outcross with some exotic mating. Each and every mating must be planned and if in-breeding is at all advised, then with the utmost caution and only to fix characteristics of such distinction that they will mark fu-

[7] When I was in eastern Europe in 1958 I was told of a well-bred Arab which had been rescued from a pigsty where it had been kept since a foal. It had to be dragged through the narrow opening on its knees.

ture generations. Never to use a mare or stallion with marked faults and because there is no other purpose for it than for stud use. This is the worst reason of all for generating a foal.

For these negative reasons, and for others, including financial, strains have disappeared. But the *positive* efforts were never more apparent than in the creation of the Thoroughbred race-horse, which we shall shortly consider; in the evolution of the Hannoverian from carriage horse to show-jumper; and in the East Prussian-Trakehner breed for its incredible stamina and endurance.[8]

Abbas Pasha, Vice-King of Egypt from 1848 to 1854, talking to the Württemburger Vice-Oberstallmeister von Hügel: "Selbst wenn es Ihnen glücken sollte, in den Besitz echter Wüstenaraber zu gelangen, werden Sie nie und nimmer wirkliche Araber von diesen ziehen können, dann die typischen Eigenscahften der arabischen Rasse verliert das Pferde, sobald es aufhört die Luft der Wüste zu atmen." In English, Abbas Pasha's remarks are as follows: "Even if you were lucky enough to become the owner of a genuine desert Arabian, never at any time would you breed real Arab horses from them, because the horse loses the typical characteristics of the Arab breed the moment it ceases to breathe the air of the desert."

What is bred in the West today, whether we like it or not, is not and has no connection with, nor resemblance to, the original horse of the desert. It is a delusion if we imagine otherwise. The breed may well stand upon its own merits as a westernized horse originally of the Arab strain or strains. It may, in some ways, be an improvement upon its generative source. According to writers of works long since out of print, the desert Arabian was not noted for a particularly good temper, indeed the raw treatment meted out to her (almost exclusively mares were used) was not conducive to fostering good temper. But again this may also have something to do with the strain. When we refer to Arab, just what do we mean? Within the desert Ku-hai-lan breed, there were the following families or strains: Sak-La-Wi; U'Bai-Yan; Ham-Da-Ni; Had-Ban; Abu-U'r-Kub; Dab-Man; Jil-Fan; Ku-bai-shanu 'l U'mair; Mi-ni-ki; Mil-wah Shar-ban; Mu-waj Ham-mad; Rab-dan; Ri-shan; Sa-dan; Sam-han; Shu-wai-man; Tu-wai-san and Wad-Man Khir-san. I have used the original spellings.

The first five of the Ku-hai-lan breed were named Al-Kham-sa (the Five) and these were the aristocracy of the Arabian race of horses. The Arabs called them *asil*, fixed breed, and it is claimed that during their 1,400 years in Nedjed (*sic*) no foreign blood was even introduced.

[8] See *Flight of the East Prussian Horses*, by Daphne Machin Goodall (David and Charles, 1973).

In his book Count Wrangel produces two excellent portraits of Arabian horses and I would submit that these two horses could belong to quite different breeds. His Nedjed stallion and Negativ the Polish-bred stallion of the Siglavi (Sak-La-Wi) have characteristics in common. It therefore seems that there is no such thing as a 'pure-bred Arab'. Western artificial selective breeding has produced *at its best* a beautiful saddle horse, generally of good conformation and gait with an equable temperament. "Furthermore," says Wrangel, "the desert Arab of the *best breeding* seldom or never exceeded 145–148 Stm.: 14·2 h.h. Europeans regard this class of horse as ponies." How, then, can anyone insist that the American 16 h.h. Arab is typical or pure bred?

Count Wrangel,[9] who certainly wrote in detail about the Arab, considered that the *original* oriental horse had one factor of importance. This factor was the importance of the breed as breeding material in Europe—seventy years ago—prior to 1909!

> It possesses no other speciality than that it is the source of improvement [of other breeds]. This single characteristic puts him upon a sufficiently high plane, that if he should disappear, it would be such a loss to horse breeding and the consequences untold. Therefore let a friendly power hold a protecting hand over the noble Nedjedi! His time is past but it could come again.

For the Western Arab his time is in full bloom. Perhaps the blood of the desert horse runs in his veins but that blood is a long way, generation-wise, from Nedjedi blood and *his* nostrils have never smelt desert air, nor have his sire and dam and their fathers' fathers. He no longer belongs to one of the *hardest* strains to be bred but—dare I say it—to one of the most pampered. The Arab is a fashionable horse.

It is, however, a fact that the earlier European studs obtained Eastern blood from Syria and both Aleppo and Damascus were markets where good horses could be bought. I find the Syrian not unlike its ancestors who drew the Hittite and Assyrian chariots. These Syrian horses, according to Wrangel, belonged in huge numbers to the Aenize (Anezah) and Shammar Bedouin. And it is from the former that the horses got their name. On the other hand the worst cross was the Al-Je-zi-ra. The fine Turkmen horses, which were 100 years ago bred to the east of the Caspian Sea and still are today, are nearly related to the Ku-hai-lan strains—or the latter is nearly related to the former! Count Wrangel states that there were three breeds: the Tekkes (Achal Teké), the Goklands, and the Yamuds (Jomuds). The first were renowned for stamina,[10] whilst the latter were famed for their speed. In Central Asia, the best of the strain standing at around 16 h.h. were called Argamak, and, crossed with the Arab, founded the Karabakh breed. Even at the turn of

[9] *Die Rassen des Pferdes*, Bd 1 and 2, (Verlag Schickhardt & Ebner, 1908–9).

[10] See *Horses of the World*, by Daphne Machin Goodall (David and Charles, 1973).

the century very few true Karabakh horses which really represented the true type of the ancient Oriental remained. They stood only 15 h.h. and had a remarkable golden tint to their coats and were called by the inhabitants Sarylar. Their coats shimmered like jewels and their manes and tails were blood-red. (See under Urartu.)

In 1858 Fitzinger refers to the Abchasen who lived on the Black Sea Coast and to their horses as the basic "strain of the Tscherkessian breed and/or as the domesticated wild rough generally white Oriental horse: *Equus caballus hirsutes*, which has changed only slightly through domestication." In 1909, Wrangel writes of them that the Abchasen horse is marked by the nobility of his appearance. The head is light and clean and often the profile slightly convex—ram's nose! They rarely exceed 14·2 h.h. and they are generally grey. They are excellent gallopers and have the extraordinary inborn ability in pitch dark to find and keep to the trickiest mountain paths. They were especially sought after by the Don Cossacks. Possibly we have here the descendants of those white horses which were so valued in the second millennium B.C.

Certainly the blood or genes of Eastern-bred horses has influenced a number of breeds of modern horses, but such 'improvement' should never be overdone in that it eventually swamps the native element. For this of itself is most valuable. The native breed characteristics, formed over generations under local conditions, can be made inefficient by overdoses of Arab blood. Also it should be remembered that native hardiness and the characteristics which make or have made a native breed worthwhile in its own right can also be improved with better living conditions.

The art of breeding good horses is based upon the happy knack of the breeder to 'nick' the best and desired characteristics of either both breeds or, individually, of both parents. Such was the case with the Orloff trotter.

One genius at specialized breeding who immediately comes to mind is Count A. G. Orloff-Tschesmenski whose stud at Khränovoi was to produce one of the world's most famous breeds of trotting horses. Orloff decided to breed strong trotters to cover the vast distances in Russia and also to race. At his age and weight it was the only participant sport left to him. He therefore bought the silver-grey Arab Smetanka who was to become the founder sire of the Orloff breed of horses. He bought him in 1775 in Moreo, for 60,000 Asignat roubles (paper money issued by the French revolutionary government). This stallion stood only 15 h.h. and was remarkable apparently because he had nineteen pairs of ribs, a fact discovered after his death. He died after only one year at stud leaving four sons and one daughter. Some writers, including Count Wrangel, insist that Orloff had a preference for motley mixtures of blood, but in my opinion his choice of matings had more than a touch of genius.

Smetanka was mated to a big grey Danish mare (Fredericksborger) and produced Polkan born in 1778. Polkan was mated to a big Dutch Hard-draver mare and produced the renowned Bars I. The Eastern-bred Smetanka possessed stamina and was crossed into a line carrying genes for high action plus stamina plus unknown. Polkan was again crossed with a big mare carrying trotting genes plus stamina plus unknown. Thus Bars I inherited a double dose of trotting genes, genes which developed high action and activity, three doses of stamina and, being bred out of a big mare with a big grand-dam was himself a big horse—he stood 17 h.h.

Smetanka was a true desert horse with the stamina and hardiness of his desert ancestors. Both the Danish and Dutch mares carried similar genes for size and for action and their Spanish forebears, the Ginetes from whom they got a definite percentage of action, were not unrelated to Smetanka stamina-wise through *their* Eastern blood. Bars I was the prototype Orloff trotter and fulfilled his breeder's hopes as early as the third generation. It is said that his son Ljubesnor I represented the ideal and perfect trotter type.

Three of Smetanka's sons remained in Khränovoi. They were Völkersalm, who left 7 sons and 59 daughters, Lubimetz who left none, and Polkan 7 sons and 21 daughters. Mated with an English mare Elawnoi, the brown stallion Bowka was bred and sold to England and he appears in the Racing Calendar of 1789 and stood at Southall. His fee of ten guineas was considered extremely high for those days but apparently his beauty, bone and action made him popular with breeders. Count Orloff also imported to Russia 22 stallions and 53 mares of Thoroughbred or near Thoroughbred blood, as the breed was just becoming established. Bars I was bred to a number of Friesian mares and three founder sires, Liouberny I, Lebed I, and Dobry I, were bred. Bars I was at stud for seventeen years. In the early days the Orloff did not cross too well with outside breeds, although now the modern Métis trotter carries Orloff blood.

In the nineteenth century severe rules were laid down for trotters and although it is not my intention to repeat them all here, it is not out of place to give some of them since they can bear upon any breed or cross. Both sire and dam have the same influence upon their progeny, i.e. chromosome inheritance is divided equally. Up to 1903, any horse with seven-eighths Orloff blood was recognized as an Orloff. If one owned a horse with seven-eighths Orloff blood the presence of one-eighth other blood was ignored. Thus if a horse had seven-eighths American Standardbred blood and one-eighth Orloff, it was naturally an American trotter. So a horse can be regarded as a cross only when not less than one-quarter Orloff blood flows in his veins. At the turn of the century there were 1,500 Trotter studs and of these 1,444 used only Orloff stallions. The remaining were American Standardbred Trotters.

Even then, the crosses were apparently faster in trotting races than the pure Orloff. But the difficulty arose in selling unwanted cross-bred horses. The Russian burghers preferred their pure-bred Orloff for general purpose use. There is a possibility that it was more even-tempered and it certainly had a stronger constitution and stamina combined with majesty of outlook which the others lacked.

The Orloff-Rostopchiner was not in any way bred to the Orloff Trotter but was, in fact, an Anglo-Arab (long before the breed was recognized in the West) of the best saddle horse type. "If beauty weighed heavier than stamina, then this horse was valued everywhere", says Count Wrangel. Practically it was for show and not for use. As always happens with such breeds, it disappeared in a comparatively short time. Capt. M. Hayes wrote:[11] "The Orlowo-Rostopchin saddle horses at Doubrovka did not please me; for they were essentially light harness animals. I would not call a horse a saddle horse, let alone a hunter, unless he had long flat oblique shoulders, light forehand, blood legs with plenty of bone, and strong hocks." Indeed, if he did not possess the latter two qualities he would not be much of a harness horse either!

Not only the idea of knighthood which originated with the chariot warriors, as we have seen, came from the East but also the ancestors of the horses upon which the knights of the Middle Ages were to ride. Strangely enough, tap root stock deriving from similar equine forebears was to provide the aristocratic descendants of the knights with a sport and a form of amusement which was to be enthusiastically received throughout the world, and the breeding of Thoroughbred racehorses was to become a major industry in many countries. The subsequent organization and running of race-courses together with the betting organization was also to become of international importance.

At first, racing, like the tourney, was a match: one horse and rider against the other. Very soon it reverted to an older idea when numbers of chariots raced against one another and carried 'colours'. From this 2,000-year-old practice, which had undergone a slight transformation in the use of colour which represented the knight's method of identifying himself, we come to the owners' colours worn by jockeys on the race-courses of the world.

Horse-racing has become the peaceful development of rivalry. It is a sport in which both the aristocracy and the ordinary people may share equally. In the same way political rivals and people of different nationalities and ideologies become sportsmen and equals once their horses are on the race-course. Monarchs mix with plain men, politicians

[11] *Among Horses in Russia*, 1900.

91. Phenomena, a Norfolk Roadster celebrated for endurance and stamina (1813 when the horse was 25 years old)

92. A common sight a hundred years ago. A Roadster could trot sixteen miles in an hour with ease

93. The celebrated trotting horse Confidence, painted by J. F. Herring, Senior, 1842

94. Shepherd F. Knapp, a famous Norfolk Roadster exported to the U.S.A., seen here running a match with the pony mare Morning Star (nineteenth century)

with their adversaries, for in the collecting or parade ring all horses are equal and they at least are unaware of rivalry or enmity.

For the moment we must leave the continental breeds of Europe and Asia although we shall later return to examine the creation of the more important light harness and saddle breeds of horses.

The Norfolk Roadster or Trotter of the British Isles was one of the finest breeds ever to be generated. When we are considering this breed, it should not be forgotten that the great Celtic tribe of the Iceni (*c.* 50 B.C.) also bred chariot ponies of international repute in the same locality as the Norfolk Roadster and within a day's drive of Newmarket, home of the Thoroughbred horse. Many coins of the Iceni period bear the representation of a well-bred pony. But the Norfolk Roadster was a treble-purpose animal, rather like the old Suffolk cob and that of Lincolnshire too. These strong, well-made hard horses were expected to work on the farm, trot to market in a Norfolk cart or dogcart, or, just as efficiently, carry the farmer himself, and in earlier days his wife as pillion passenger, wherever they required to journey.

Like the description of Clement Paston of Paston, Norfolk, in 1350: "He was a good plain husbandman, and rode to mill with his corn under him, and brought home the meal again under him."

Later on, if this splendid all-round cob possessed just that little bit extra, then he or she, for the mare Phenomena was famous, would be raced in trotting matches.

The early historic origin of the strains mentioned above would appear to lie in the Fens, whence domesticated (chariot) ponies would have been driven to avoid capture by Dane and Saxon invaders after the Roman withdrawal. Both Danes and Saxons relied upon local horses to help them foray further inland. In an attempt at survival, it is recorded that the Fen ponies had developed large heads and feet but were not above 14 h.h.

In 1171, the monk William Fitzstephen or Stephenson wrote an account of the City of London which he called *Descriptio Civitalis Londiniae*. It is interesting from a horse-breeding angle since he describes the Friday horse-market at Smithfield. An exact account is to be found in our book *The Foals of Epona* which deals so specifically with the breeds of horses and ponies in these islands that I shall only briefly, in so far as it affects this history, go into them here.

The first horses which Fitzstephen mentions are palfreys, which most probably were imported (from Spain) and were the ancestors of the Ginetes. These horses, really not much bigger than ponies, ambled, a gait which made them exceedingly comfortable for long journeys. Then come "horses more suitable for squires with rougher but faster gait". These were trotting cobs. There follow pack horses with stout and nimble legs, and "Expensive chargers of handsome conformation and impressive height." "Mares for ploughing or draught in sled or double harness cart." Also mares in-foal or with foals at foot.

In an earlier chapter, I wrote that high action appeared to be synonymous for an ancient Iberian strain, such strains inhabiting in prehistoric times certain river valleys. In the same way we now find that in the England of the Middle Ages, the trotting gait appeared to be a peculiar characteristic of horses bred on marshland, for this gait was peculiar *at that time* to horses coming from the eastern regions of England. In Chaucer's *Canterbury Tales*, the Reeve, a Norfolk man, rides a good trotting mare, a dapple grey cob in fact. They were called *runcini*, and the Domesday Book (1086) lists a very high proportion of *runcini* on some Norfolk manors. "Elizabeth I's order for mustering Dymlances and Light Horsemen enacted that every dymlance should be mounted on a trotting stallion or a trotting gelding."[12] Amongst the trotting strains were the now extinct Lincolnshire Fen pony from which was descended the Lincolnshire Trotter, the Norfolk Roadster, which, being a fixed breed of many generations, improved so many other breeds, the Suffolk Punch, originally a trotting cob, and the modern Hackney.

Since mail was always carried on horseback, fast trotting horses capable of carrying a man travelling long distances were much in demand. The early trotting matches in Great Britain as well as Holland were also ridden and later driven. A speed—but not the distance—of sixteen miles per hour was normal. Phenomena, who stood only 14½ h.h., was matched in June 1800, at twelve years of age, to trot 17 miles within the hour, in this case the distance was included, on the Huntingdon road. She performed this with ease in 50 minutes; a month later she completed the distance in 53 minutes. During this match she did 4 miles in under 11 minutes, timed by stop-watch.

Phenomena bred only one descendant, as far as we know, by the blood horse Walton (1799) who had Regulus blood both sire and dam. Regulus appears in a number of coach horse pedigrees. This offspring, a bay mare, bred Jacob's Phenomenon who sired Norfolk Phenomenon (Taylor's). Phenomena was ridden in her matches. Sorrel or red roan was the colour common to both the Suffolk (Punch) and the Norfolk Cob (Roadster), of the very early 'old breed', to which many references are made in old documents. It generally produces the chestnut colour. Sorrel is also the colour of the Schleswig, Jutland and Ardennes breeds of horses. The colour is not peculiar to any other British breed.

There was another great trotter mare, of whose prowess records have been kept. This was Nonpareil, who covered 100 miles in 10 hours, 14 minutes and 40 seconds, including stoppages, in harness over Sunbury

[12] *The Foals of Epona* by Daphne Machin Goodall and A. A. Dent (Galley Press, 1962).

Common on the night of 9 June 1835. She had been allowed ten and a half hours, and she was, it was reported, "fresh to the last".

During the late eighteenth and early nineteenth centuries, these trotters were an exceptionally fine breed of horses of enormous stamina and endurance by sires such as Grey Shales, Marshland Shales, Wildfire, Fireaway, Young Prickwillow, Don Carlos and North Star. A Prickwillow appears in the pedigrees of some Hannoverian horses; whilst North Star went to Hungary to found a line, the North Star-Furioso. Like the Spanish horse of a few centuries earlier, so many Roadsters were exported that none remained to carry on the breed, although some blood flows in the veins of the modern Hackney. Here again, however, few of the old genuine large Hackney horses are to be seen today.

In 1829, three foreign dealers bought around 300 mares at Horncastle Fair and year after year at all the sales and fairs the best mares left the country. In 1844, Mr Richard Tattersall said that America and the countries of Europe had been buying the best stallions and mares that money could buy in England during the last hundred years and more. When it was too late, steps were apparently taken to try to retain and revive a semblance of the old trotting strain.

A correspondent in Holland informed me that in 1864, in the Friesian village of Winsum, a British firm acting for Yorkshire and Suffolk breeders appointed a representative to buy for three years pure Friesian horses of the best type. In that year, it is said, 100 were sent to England, in 1865 79, and in 1866 157. Then in 1877, the same reliable source—a dissertation held in the University of Utrecht—states that breeders of the Groningen warm-blood horse imported several black Norfolk and Suffolk stallions[13] with a view to improving the Groningen breed, a vanner or very light draught horse type. In 1885 a jet-black, *gitzwart* Norfolk stallion was declared champion and best of all breeds at a great exhibition in Amsterdam. This horse belonged to the Grand Duke of Luxemburg and had been presented to him by King William III.

Thus, only one hundred years ago there was a regular exchange of blood between Friesian, Yorkshire, Norfolk and Suffolk breeders. This lends credible support to the opinion that much earlier Spanish blood, possibly in part from Holland too, found its way into English East Coast breeds. Now in 1974 only the Cleveland Bay remains on its native pastures, and the Hackney in studs in Europe and the United States, but it is no longer a breed native to any particular country.

But success in reviving the old Trotter was only temporary since today it is very rare to find a trotting cob in East Anglia. During the 1930s my sister owned and show-jumped a Norfolk Roadster called YNot? This was a chestnut gelding of tremendous character, stamina and ability with a

[13] Although black was common to the Lincolnshire Trotter, I have not before come across it especially in these breeds.

profile slightly convex. I have owned a brown cob mare standing only 14 h.h. with $8\frac{1}{4}$ inches of bone, with a similar slightly convex profile—like the Nisaean she was fast, active, strong and willing, and, like them too, she was inclined to be 'untrue' in the stable!

The Norfolk Roadster was exported to France and helped to improve the Norman breed; and together with Messenger by the grey Mambrino, Bellfounder, another Norfolk Trotter, founded the American Standardbred breed of trotters. First amongst these was the famous Hambletonian to whom many modern Standardbreds trace their ancestry.

The Norman crossed with the Standardbred founded a new breed of trotters: the Noram which is very popular in Europe. The Orloff crossed with the Standardbred also fathered a modern strain: the Métis trotter.

The Roadster in Great Britain eventually evolved into the high-stepping active Hackney which is known throughout Europe, the United States and Canada. As yet, however, the Hackney has had no influence on the evolution or creation of newer breeds, but where Hackney blood has been used, the resulting foals more often than not make top-class jumpers. It is unfortunate that breeding of horses with the courage and ability to jump has not been encouraged and developed in England.

There is one other breed, native to Yorkshire, which also had an infusion of Spanish blood and which in turn helped to found the present breed of Hannoverian horses. This is the Cleveland Bay, which thanks to the enormous care bestowed upon it by the breed's admirers in the Cleveland district of Yorkshire is happily still alive today. It nearly vanished—but not quite!

As far as we are able to trace, the Cleveland Bay, originally very much a locally bred animal, was partly descended from the Chapman pack-horse, sturdy, short-legged horses which the chapmen (travelling salesmen) used to bring wares and merchandise from the north to London.

The heavier type of Cleveland Bay was used for draught and agriculture, whilst the lighter type made first-class coach horses once the roads were fit to travel. Crossed with Thoroughbred horses, the strain became known as the Yorkshire Coach horse. In general the Cleveland Bay, like the Irish Draught horse (the heavier Cleveland Bay and the larger Irish Draught horse are not unlike) when crossed with the Thoroughbred produces both hunters and show-jumpers. It is not a cross, however, in which further evolution seems possible. In other words, such a cross cannot evolve into a breed since breeders wish to retain the true Cleveland Bay type and strain and to do this one must breed back to the original strain, the Cleveland Bay.

As we have already seen, horse breeding in Saxony was general from early times. The first stud to be directly concerned with laying the foundations for the future breed of Hannoverian horses was Bücken, District Hoya and Memsen. In 1735, Celle was chosen as the headquarters for a stallion depôt. At first too many horses of doubtful descent were used. The preceding chapters have shown that there were very few fixed breeds even at this time, and, as a consequence and as could be expected, horses of very uneven type were bred.

About one hundred years later, a number of Thoroughbred horses were imported from England and although there was an improvement in quality, the use of Thoroughbred blood was overdone and a light race-horse was the result. This type was more or less useless in a country where carriage and coach horses were increasingly in great demand. To counterbalance, the breeding lines of the mares were 'strengthened' by Norfolk Roadster, Cleveland Bay and Yorkshire Coach horse blood.

It was clearly important to lay the foundations for a native Hannoverian breed—a quality half-bred warm-blood horse. The chief breeding centres were in the river marshlands of the Elbe, Oste, Aller, Weser and Leine and breeding was and still is in the hands of farmers and land-owners.

When, at last, there were a sufficient number of representative stallions, 170 in all, a stud book was formed in 1893. The Hannoverian horse was a type between the heavy Oldenburg coach horse and the elegant East Prussian saddle horse. Horses with good action and temperament were required. These, incidentally, are two of the characteristics of well-bred horses, with fixed blood lines. (Bad action and bad temper or unreliability are found in horses which have the misfortune to be the result of a bad cross!)

The convex profile and sometimes the straight shoulder of the harness horse were also at one time a Hannoverian characteristic. The convex profile is not surprising since some of the original founder sires were, as we know, Spanish and Neapolitan. Both these characteristics, undesirable in a well-made saddle horse, have now been bred out, whilst generally the ability and courage to jump have been retained, for the modern Hannoverian is an internationally recognized show-jumper.

The two main stallion depôts are at Celle and Osnabrück whilst the two chief breeding centres for the colts are Emsland and Artland. Colts are selected and taken to the first-named centres to be trained and schooled to the saddle and in harness. Tests include cross-country and draught in a weighted sled on runners and are so arranged that the inspectors can determine the entire's full potential, health, character and temperament as a future sire. When the selectors, who include the Stud Director, inspectors and veterinary surgeons, are fully satisfied with the horse's performance, a certificate will be awarded. The horse will then, with his groom, be allocated to a district. The suitability of the stallion as a partner for the mares of that district is also carefully considered.

Between seasons the entires are stabled at Celle, once the seat of the Electors of Hannover, and every autumn about 200 horses give an exhibition. Free dressage, jumping, musical rides, trotting to sulkies, and teams of up to twelve horses to the royal state coaches show the achievements in training of which Hannoverian stallions are capable.

During the season for 1973, 11,645 mares were covered, showing an increase of 7 per cent on the previous season. Thoroughbred and Hannoverian stallions were used. As long as the mare is Hannoverian, e.g. entered in the stud book, her offspring are Hannoverian even though the sire may be Trakehner or Thoroughbred.

Almost all the European breeds of warm-blood horses were created either to serve agriculture or to furnish remounts for the army as cavalry or artillery horses.

It is seldom realized what a vast number of horses were used in the decades that followed the establishment of the various breeds. Nor, indeed, is it generally appreciated that well over one million horses lost their lives in the two world wars.

In 1732 the Stud Trakehnen[14] in East Prussia was founded by Frederick William I. Over 1,000 mares and stallions of mixed strains and origins and very uneven in type were at first used as breeding stock. Seven years later, the King gave the entire stud and domain consisting of some 6,000 hectares (14,000 acres) to the crown prince, later Frederick the Great. But breeding was continued on a rather haphazard and amateur basis. On the death of the King in 1786, the stud passed out of private royal ownership and was taken over by the Prussian state. Breeding stock was thoroughly sorted out and many horses were culled. The mares that were retained were divided into saddle and carriage horses and into five herds for colour: black; brown/bay; chestnut; and two mixed—grey, dapple grey and other colours. The three single-colour herds were for carriage horses—since pairs and teams should be matched—and the mixed colours for saddle horses. The lighter type of mares were kept in Trakehnen and the heavy type at Bajohrgallen. Probably the quality of the soil and grazing led to this decision, since both influence the development of horses and cattle. The soil generally in East Prussia varies from medium to light sandy.

Trakehnen was to become virtually the 'Newmarket' of the East Prussian warm-blood horse. From 1787 all horses born in Trakehnen were branded with a single seven-pointed elk antler brand on the off-thigh. From 1888 the ordinary East Prussian horse had a double antler brand on the near thigh and it is still the practice to brand the horses which have found a new home in studs in West Germany, to which they came as refugees. Oriental and Thoroughbred sires were used, the most famous being Perfectionist, son of Persimmon, owned by King Edward VII. He

[14] See *Flight of the East Prussian Horses* by Daphne Machin Goodall for a complete history of the breed.

95. Hackney pony Little Ruby, born 1917, died 1945

96. Sable Island ponies, Nova Scotia. These ponies are feral, and have been on this Atlantic island for about 300 years. Similar ponies were taken to Canada and North America at the same period

was three years at stud (suffering a fatal accident in his box), but left behind 32 stallions and 37 brood mares.

Trakehnen was evacuated four times during the course of its duration, for the last time to which there was to be no return, in the winter of 1944–45 in the face of rapidly advancing Soviet troops. Between the wars the province of East Prussia, one of the most beautiful countries in all Europe and perhaps in all the world with its vast forests, lakes and wild life, exported 32,000 horses annually. In the flight only 700 East Prussian mares and a few droves of colts and stallions reached the West. The mares were in-foal and unshod, some were scarcely ever taken out of harness in two and a half months. They covered 900 miles, pulling wooden field waggons with their owners and their few possessions through snow, ice, and gunfire over cart tracks and muddy 'summer' roads,[15] in just under three months. Many died or were killed, most survived on what food could be found in a devastated countryside.

One of the refugee owners wrote: "The endurance of our East Prussian horses during the flight from our native land to the west is unparalleled. One can only hope and wish that in the re-building of our country an appropriate place will be found for them."[16] Unhappily these horses came to the west of Germany which was already faced with the problems of the aftermath of war, occupied by three foreign armies, overcrowded with refugees (13 million Germans were made homeless and had been forced to flee), and with its own native breeds of horses.

The Trakehner horse had been used, however, to improve these warm-blood Western breeds. It had been bred for stamina; one might almost say that the entire *history* of the breed had led to this test of survival. And survive it did, in spite of almost overwhelming odds. The breed intact is to be found in Schleswig-Holstein, in the Solling, and in many private studs.

East Prussian horses are again showing the elk-antler brand, not only in international events but in many countries throughout the world where studs have been formed!

Also in Masuria, which was handed over to Poland after the war, the East Prussian horse once more grazes in its native land. Trakehnen itself was practically destroyed and under Soviet occupation was used, so I was told, to breed heavy horses and cattle. I have been invited by the Polish Ministry of Agriculture twice to visit officially studs throughout Poland and East Prussia—or Masuria as it is now called. The Poles who now occupy the western half of the province have rehabilitated the studs and stallion depôts. In place of the East Prussian, the Masurian horse is bred

[15] In Eastern Europe all tarmac roads have ordinary earth tracks alongside, especially for horse-drawn vehicles.

[16] From letters in the possession of the Society of Breeders and the Friends of the Warm-blood horse Trakehnen Breed.

with the same old blood lines and similar pedigrees as its contemporaries in the West.

Poland belongs to the classical horse-breeding countries and the Poles themselves are real horse lovers. Not 400 years ago in Poland, the smallest stable consisted of 100–200 horses. When the aristocracy travelled at least 500–600 horses were necessary to carry their court personnel and belongings from one place to another. And there were too, in Poland, Lithuania and in Prussia many wild studs, the term here meaning in many instances, really wild Tarpan; whilst in others the horses were kept under the same conditions as wild herds.

The states were vast. One can travel, as I have done in 1958 and 1964, through the unending countryside, sometimes driving for fifty miles without seeing a soul. The second war produced many difficulties, not the least being that the large estates and private studs were destroyed. There were, it is said, 3·6 million horses in Poland before the war and by 1961, in spite of the material losses, the number had again risen to 3 million.

The Polish State Stud directorate is the largest in the world with 42 state studs covering 88,000 hectares with 3,300 brood mares and 16 stallion depots with 2,500 entires at stud. This includes the evacuated and neglected German studs in 1945 at the end of the war. At one time, in the state studs there were 800 East Prussian mares and 900 Posen mares. Many of these had been found abandoned, victims of the aftermath of battles which had devastated the farms or studs where they belonged. Many, too, were without names or pedigrees or papers, identified only by the brand. Later a few were re-identified, as were a few of the abandoned stallions. In fact, I photographed one in 1958, showed the photograph to breeders in the West, and his identity was established! On this basis the Poles have built up their Masurian and Poznan breeds, called I believe Wielkopolska.

A number of horses were exported annually to Switzerland as remounts, but since Swiss policy of recent years has been to cut down her cavalry units, possibly this source of export has now ceased.

Poland is, perhaps, most famous for her Arab studs of the eighteenth and nineteenth centuries: Sawran, Guminska, Slawuts-Chrestowka (the two latter belonging to Prince Sanguszko), Antoniny (which belonged to the Potocki family who also bred odd-coloured horses), Jarezowce, Jezupol, Jablonow, Jarzeze, Taurow and Pelkine (which belonged to Prince Czartoryski). Many of these nobles imported valuable Arab horses from Asia and especially Syria, the most intrepid and legendary being Count Waclaw Rzewuski who went himself to Damascus and resided there during his search for first-class horses. The first of these found their way to Weil at the request of the Queen of Württemberg. According to Ammon, this stud "had no equal in the whole of Europe". Many other

97. *(top)* Arab mares at the Janow Podlaski Stud, Poland

98. *(right)* The roan Haiti horse, whose profile bears a resemblance to the Sorraia. These horses are wild in the high altitude pine forests of Haiti, and this one was shot in 1939. Skin and skeleton are now in the Natural History Museum, London

99. *(above)* Speedy Scot, a Standardbred American trotter, painted by Elizabeth Bell

good horses bought by Rzewuski were sent to Poland.

During the eighteenth century there were frequently wars between European countries and the Turks. Hungary was under Turkish dominion for 160 years. By this means vast numbers of Oriental stallions came into these countries. After one such war according to Rzewuski, 800 entires came to Poland and the Ukraine.

The studs in Russia were very numerous too and Arab stallions were imported at considerable cost by the wealthy nobles. Count Sangusko sent his stud manager to Syria where he bought six expensive and valuable Arab stallions. By the nineteenth century, however, in European Turkey the quality of horse breeding had decreased to such an extent that only very moderate horses could be obtained and they were by no means pure bred, being a mixture of the old Thracian, Bulgarian and Greek breeds together with Eastern Breeds. And these bastardized crosses were regularly imported into Europe!

Now it was apparent that the necessity to breed horses of similar conformation and characteristics was becoming very real. And it was out of this necessity that the modern European breeds evolved by careful selective breeding.

Janow Podlaski on the River Bug, near Brest, is today one of the largest studs for pure-bred Arabs and half-bred Arabs in Europe. This stud, which lies 200 kilometres from Warsaw, was founded in 1817 by Czar Alexander I with 100 mares and 55 stallions acquired from private studs. The first Arab stallions were sent by Count Rzewuski in 1822. At first the stud bred Thoroughbred horses which were later disbanded. For some reason the Thoroughbred has never done especially well on Polish soil. Pure-bred Arabs, half-bred Arabs and Anglo-Arabs were also bred there.

The Irish Draught horse, whose origins are rather obscure, was in part distantly related to a contemporary—the Spanish horse whose breeding we have examined elsewhere. The obscurity is due to the lack of records. The earliest entries in the stud book bear Dutch names but whether the owners were really descended from Friesian or Spanish imported horses is impossible to say.

Volume I of the stud book informs us that as a result of the close association from earliest times between Spain and Ireland, many horses were imported and these were presumably Spanish (Andalusian). As well, there could have been some imported Villanos. It seems quite within reason to suppose, failing other evidence, that this could have been the source for the heavier frame and bone.

The Connemara[17] pony, of course, was native to Ireland from early historical times and being a mountain and moorland breed it was tough,

[17] See *The Foals of Epona* by Daphne Machin Goodall and A. A. Dent.

hardy and strong but not strong enough, I think, alone to have produced the *bone* of later Irish Draught horses, which incidentally had a distinctly cobby outlook[18] and were shorter on the leg than they appear to be today. The renowned chalky soil of the Emerald Isle would account for an increase in bone and the necessity for sturdy pack-horses would lead to a kind of selective breeding.

The sad thing is that the Irish themselves took the science of breeding so unseriously that the Irish Draught horse was at the beginning a very haphazard product. But in my opinion the influence of the Spanish horse is not to be denied in the activity, action and convex profile of this extraordinarily useful animal. The first cross of the Irish Draught × Thoroughbred—a back nicking of genes?—produced the best hunter in the world, a horse of courage, stamina and jumping ability. These characteristics later proved of immense value when show jumpers became popular. However, whilst the first cross was excellent, second and third crosses are said not to be so good. The Irish Draught horse is not and never was a cart horse and has no connection with heavy draught breeds. It is a coach horse type standing around 16·2 h.h., has good bone and joints, whilst grey used to be the favourite colour.

Before the First World war and between the wars, horses were shipped to Ireland from East Prussia. There was a certain scandal when it became known that yearlings of the Hannoverian breed were not being branded, in other words they were not declared to the Stud Book Commission for inspection and entry in the German stud book. Free of a brand they were then shipped to Ireland. One wonders how many Irish hunters were actually bred on the Continent!

There is no reason against the practice of *raising* young horses, not bred in Ireland, on Irish pastures and perhaps it is splitting hairs if a yearling of non-Irish parentage after four seasons' grazing there is called 'Irish'. But it does account for the impossibility of organizing a stud book for warm-blood horses and for the fact that the famous Irish hunter and jumper is not a breed as it might so well have been. The matter of the stud book has now, I believe, been corrected.

It is vigorously maintained that there were no horses in either North or South America when the Spanish *conquistadores* appeared during the sixteenth century. This may be so. Experience has taught me to be a little sceptical when such all-embracing statements are made. The two continents are vast in extent and new discoveries appear to indicate that that strange horse with the large head, *Onohippidon*, was roaming the South American pampas when the horses from Europe appeared. These discoveries of fossil bones and hooves seem to indicate that the extinction of this species happened within recent times.

[18] *Ein Leben mit Pferden* by Graf Lehndorff.

Thus the existence of the Haiti horse still poses a mystery. Is it an ancient form or is it a descendant of those Spanish invaders of nearly 450 years ago? The difficulty is to discover not the strains which the Spaniards took with them but how many horses were landed in the two Americas and the islands of the Caribbean! It has been hinted to me that in the libraries of Mexico City a great quantity of unexplored and therefore unpublished material is to be found.

Bernal Diaz de Castillo who became Governor of Guatemala wrote as an old man, a history of the Conquest, and in so far as we are interested in the horses, here is his description from memory:[19]

> Captain Cortés had a dark chestnut stallion which died when we reached San Juan Ulua.
>
> Pedro de Alvarado and Hernando López de Avila had a very good sorrel mare, turning out good both for tilting and for racing.
>
> Alonzo Hernandez Puertocarrero had a swift grey mare which Cortés bought for him.
>
> Juan Velásquez de León also had a sturdy grey mare which was called La Rabona on account of her docked tail. She was fast and well trained.
>
> Cristoval de Olid had a dark brown horse.
>
> Francisco de Morla had a dark brown stallion which was fast and well-reined.
>
> Francisco de Montejo and Alonzo de Avila had a dark sorrel which was useless for war.
>
> Juan de Escalante had a light bay horse with three white stockings but she was not very good.
>
> Diego de Ordas had a barren grey mare, a pacer, who therefore seldom ran.
>
> Gonzalo Dominguez, who was an excellent horseman, had a dark brown horse that was spotted. He was worthless.
>
> Lares, a good horseman, had a very good bay horse who was a good runner.
>
> Ortiz, the musician, and Bartolomé García, who had gold mines, had a black horse called El Arriero—which means The Mulateer [he had probably driven a pack train, and he was one of the best horses taken in the fleet].
>
> Juan Sedeno had a brown mare that foaled on board ship.
>
> These were the sixteen horses which were embarked for Mexico.

In fact there are only fourteen including the foal!

The Indians are said to have observed to their Spanish conquerors: "You bring with you those animals which our fathers knew." This may have referred to a recent memory of horses present in the Americas or it may have referred to a traditional memory going back several thousand years, since the Indians are of Asiatic origin of European and Mongolid ancestry. Their way of life, at least, until the newcomers helped to destroy it, had many things in common with the way of life of those

[19] Diaz, *Historia Verdadera de la Conquista de la Nueva España.*

steppe nomads who appeared in the early chapters of this book. It was a continuity of more than two thousand years—perhaps even much longer. The Indians very rapidly proved themselves to be *born* horsemen.

We do not know the colour of *Onohippidon* but the colour, red roan/sorrel (there were two sorrels in the above list) of the Haiti horse is a primitive colour, and the primitive Sorraia is roan or dun. The head and profile of the Haiti horse can be compared with the head and profile of some of the semi-wild Sorraia horses. There is, in my opinion, a distinct resemblance but this opinion neither admits nor denies the authenticity of the Haiti horse. It may represent an ancient prehistorical species, or it may—like the ponies of Sable Island in the Atlantic—represent a living strain of 450 years ago, a strain that has had no crosses of outside blood for so many decades.

The Criollo of South America is certainly of Sorraia-Garrano-Spanish descent. One is reminded of the American cowboy songs about the Zebra Dun and the Strawberry Roan.[20]

There are no less than ninety-five registered Criollo coat colours in the Register de la Sociedad Rural Argentina. The horses vary little in conformation from Brazil to Chile and most have a profile that is either straight or slightly convex. But although they share similar characteristics of conformation, they do seem to vary in weight, the heavier type suggesting Villanos blood in some of the original horses brought to the New World. The heavier Criollo is not at all unlike the cob which was once common although perhaps not plentiful in Europe; only here and there is it mentioned by earlier writers. The Criollo, like the cob or its Spanish ancestors, is an all-round horse, ideal for riding over long distances, as a cow-pony and in harness.

In the United States of America there are now a number of breeds and colour strains which owe their origin to ancestors imported during the sixteenth century. Descendants of these imported horses were cross-bred with later imported horses from Europe such as the Andalusian and the Norfolk Roadster, Thoroughbred, and of course the Arab. Breeds such as the Morgan, Quarter Horse (which was a *quarter*-mile sprinter) Naragansett Pacer, Tennessee Walking Horse and the American Saddle horse have been selectively created within the past 200 years. The period of their existence is, however, too short for these breeds to be properly considered within the confines of history.

At the same time that these first-class saddle horses were being created during the late eighteenth century, the first Thoroughbred race-horses were also taken from Great Britain to America. The oldest race-course in the United States was laid down in 1827, but the first attempt at a stud

[20] *Mustangs and Cowhorses* by J. Frank Dobie, Mody C. Bodright and Harry H. Ranson.

book, Wallace's American Stud Book Volume 1, made its appearance in 1867.

Many race-tracks are on sand and this encourages breeding sprinters, as indeed does the policy of early maturity. There is a vast race-going public. Some unbeatable horses like Ribot have been bred, but, as with the saddle horse breeds, it remains to be seen what influence these great American-bred Thoroughbreds—bred for speed rather than staying power—will have on future generations of racehorses or the evolution of future strains.

The American Standardbred Trotter evolved from five basic breeds: Thoroughbred, Norfolk Trotter, Arab/Berber, Morgan (which probably had some original Spanish blood), and Pacers which themselves belonged to different breeds. Although the breed is always referred to as the American Trotter, there are also Pacers of the same breed, just as there are races for Trotters and for Pacers.

In 1893 Sanders wrote: "It is not true that we have a special breed of harness horses. The horses we use for light draught, fast trotting, etc., are mainly a mixture of many breeds and types."

The point that appears to be generally accepted is that the founder of the Standardbred, whose peculiar characteristic is either pacing or trotting—and neither of these gaits is supposed to be an equine characteristic—was the imported Thoroughbred Messenger by a great grey horse Mambrino, whose sire was Engineer by Sampson by Blaze, a great-grandson of the Darley Arabian. Not that this is of so much consequence, since there is no proof whatever that a great-grandparent's genes are passed on to his descendant. True, some *may* be but also they may not be. It is an interesting speculation and one which I have never seen satisfactorily examined nor explained by science.

Flying Childers, the sire of Blaze, was described as "Generally supposed to have been the fleetest horse that was trained in this or any other country". Sampson had also the blood of the Brownlow Turk, Rutland Black Barb, Curwen Bay Barb and Lister Turk. So perhaps it was, after all, Berber blood that was to be responsible for future great trotting horses (see page 166 on the Berber).

Bellfounder, the wonderful Norfolk Roadster, was also imported to the United States in 1823 and he (as a contemporary report stated):

> was got by that well-known fast and high-performed trotter Old Bellfounder, out of Velocity by Haphazard, by Sir Peter out of Miss Hervey by Eclipse; granddam was of good north country blood, but not Thoroughbred. Velocity trotted on the Norwich road in 1806 16 miles in 1 hour and she broke fifteen times into a gallop, and as often turned round, won her match. In 1808, she trotted 28 miles in 1 hour and 47 minutes. Bellfounder at five years old trotted 2 miles in 6 minutes, and in the following year was matched for 200 guineas to trot 9 miles in 30 minutes, which he won easily by 32 seconds.

However, it is a fact that the blood of these horses combined to produce Hambletonian 10 or Rysdyk's Hambletonian, born in 1849, a great sway-backed horse whose powerful quarters appeared not to belong to the same back and forehand. His dam, by Bellfounder, had apparently so injured herself that she was about to end in the slaughter-house. She was bought by a man who had discovered that she was out of a mare called One Eye which his father had admired. On acquiring her, this crippled mare was sent to be mated to an ugly, bad-tempered stallion called Abdallah who was then twenty-eight years old—his grandfather was none other than the great Messenger. No one knows who was responsible for poor Abdallah's bad temper but in the end it was responsible for his miserable end. The following spring, 1849, the crippled mare gave birth to a rough-coated colt foal. Rysdyk, who worked for her owner, bought her from him on credit for 125 dollars—and thus both mare and foal found a home with someone who evidently had an eye for a horse. And what a horse Hambletonian turned out to be!

Poor Abdallah kicked the fishmonger's cart, to which he was harnessed, to pieces and since the fishmonger considered him no longer worth feeding, he turned him loose on the Brooklyn shores where he died of starvation.

The Standardbred Trotter possesses much courage and stamina and is generally kind. At first these horses were required as light harness and carriage horses and the first trotters appeared in their matches under the saddle, as in Holland, and in Great Britain as well; although Norfolk Roadsters were used both under the saddle and in very high light carts, the forerunners of the sulkies. Then, during the last part of last century, sulky racing became popular across the Atlantic. The horses are bred right across the United States and Canada, but Kentucky with its marvellous grass pastures is really the recognized centre for the Standardbred Trotter.

As far as breeding and racing Trotters in Germany is concerned, the sport took off comparatively late, in fact, only 100 years ago when the Hamburg-Altonaer Racing and Trotting Club was founded in 1874. At first Orloff stallions were used and also Orloff mares, although after 1942 no further Orloff mares are entered in the stud book. There is a marked division between pure-bred and half-bred Trotters. Both American and French Trotters have been introduced, and today the chief trotting lines are based on the former.

As with trotters of 200 years ago, the value of a modern Trotter rests on his record. There are large studs in north and west Germany and one- and two-mare owners amongst the farming community. And basically, as with the breeding of many warm-blood horses in Germany, it is to these farmers and the mares they keep that the breed owes a reserve of stam-

100. Lord Oxford's Dun Barb, a Berber horse, 1715

101. Blank, whose interesting pedigree has the Godolphin Arabian (or Barb), Darley Arabian, Woodstock Arabian, St Victor's Barb, and the Fenwick Barb

102. Coneyskins shows a marked Berber influence (1712)

103. The Byerley Turk, a different breed from the Lonsdale and Grosvenor Arabians, but probably of the same breed and area as Lord Oxford's Dun Barb and the Bloody Shouldered Arabian

104. The Princess, winner of the Oaks, 1844

ina, hardiness, longevity and good temperament. The Trotter mare Miss Benton covered the 750 kilometre distance between Berlin and Totis in Hungary in 110 hours 22 minutes.[21]

As I have already noted, trotters provide the best breeding material for breeding show-jumpers. One of the greatest international show-jumpers was the mare Halla by the Trotter stallion Oberst. She was retired to stud.

The creation of a new strain of racehorses, later called the Thoroughbred, before, during and after the reign of Charles II, from imported Eastern and North African sires and native and royal mares was a masterpiece of generation. It was a gradual process that took several hundred years to mature.

Since I have dealt with the early antecedents of the Thoroughbred horse in previous chapters, I do not propose detailing them further here. We take up the story of the evolution from those many strains which form the background to the generation of the world-famous Thoroughbred racehorse.

Some stallions in the early generations of the Thoroughbred influenced a number of other breeds of horses in other countries and generally to the immense advantage of that breed or intended breed. The question arises, however, if the same thing can be said today—can the sprinter so influence blood lines of the equine species, so that within a few decades a new strain would be recognized as a fixed breed with definite desired qualities of its own?

Originally these horses were bred for stamina, today they are bred for speed. So we may ask: Is today's Thoroughbred similar to that of a hundred years ago? And in a hundred, fifty or even twenty-five years, what will be the position of the Thoroughbred as the premier breed? Will breeding for speed—the automation of the horse—destroy health, and this means constitution and stamina?

For a very long time both the monarchs of Britain and the ordinary people enjoyed the sport of running horses, in early times in chariots, as we have seen, and later when one running horse was matched against the other. It was in the sixteenth century that the breeding of such horses began to take on a new significance. J. P. Hore writes:[22] "In the spring of the year 1514 Giovanni Ratto was sent by the Marquis of Mantua with a present of thoroughbred horses to Henry VIII. From these and some subsequent drafts were descended, in all probability, many of the English racehorses of the sixteenth century."

The horses were presented to the King on 20th March, "at a place

[21] *Handbuch der Tierzuchtung*, Bd. 3 (Ensminger & Uppenborn).

[22] *History of Newmarket and Annals of the Turf*, 1885.

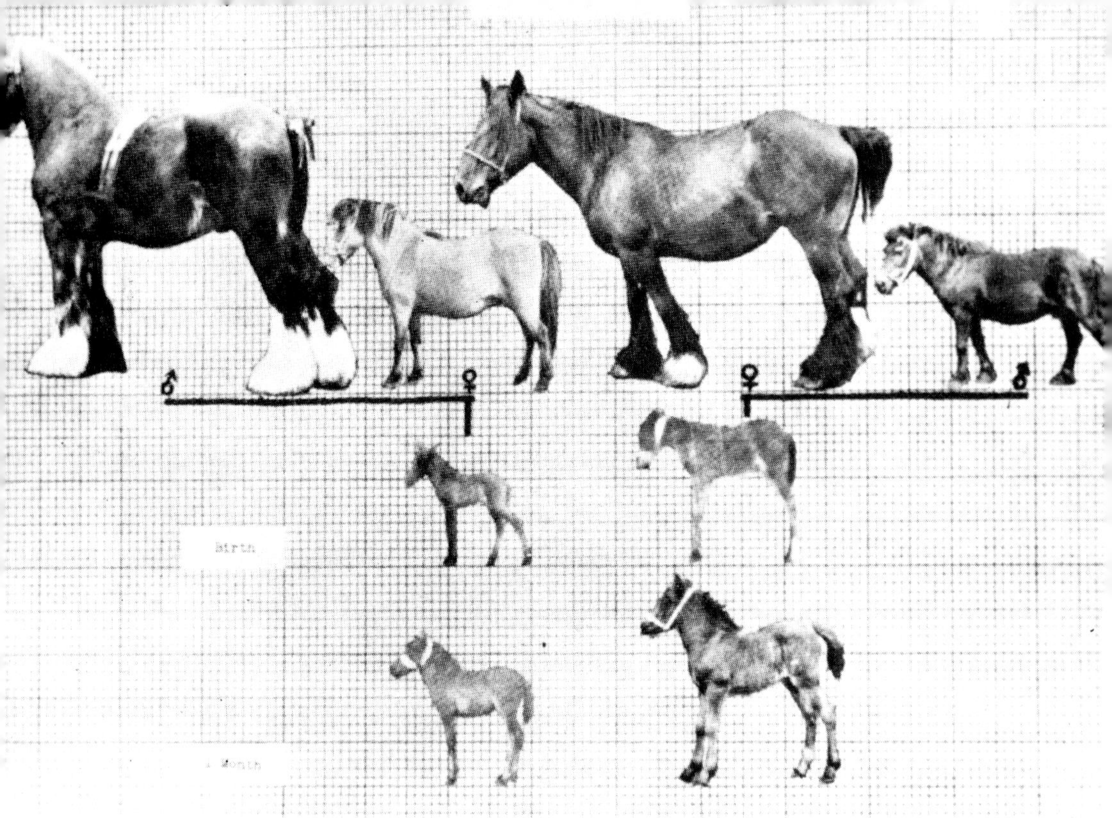

105. *(above)* Left, the progeny of a Shire stallion and Shetland pony mare, at birth and at one month. Right, the progeny of a Shetland stallion and a Shire mare, at birth and at one month

106. (*left*) Variations in the size of horses. These illustrations and those opposite are taken from *Genetics of the Horse* by permission of Dr Wm. E. Jones D.V.M., Ph.D.

107. *(above)* The cross between a jack ass and mare results in a mule, taking after the mare; *(below)* The cross between a horse and a jenny donkey results in a hinny, taking after the jenny

called Hampton, four miles from London". And the King was so delighted, he went from one nobleman to the other, asking: "What think you of these mares?" The King was astonished to see the mares in action and said he had never beheld better animals. The French Duke of Longueville told the King that his master the King of France had no such valuable mares. I think it is important that we appreciate the *value* of these Mantuan mares in the first instance, in helping to provide the foundations upon which the breed was to evolve.

Ratto further informed Henry that the Marquis had a stud of Barbary mares, of 'miche' and of ginetes, and also of great mares which he offered to the King. "The queen was present during this conversation which induced Ratto to put 'the bright bay' through his paces in the Spanish fashion", exhibiting the horse to the admiration of all present. Mention is made, according to Rawdon Brown, of this horse in the *Fioretto delle Croniche di Mantova.* He was a Mantuan 'barb' or racehorse and the Marquis had been offered his weight in silver for him, but preferred to present him to Henry VIII.

The King informed the Marquis of his delight at these presents, "as the horses were not only beautiful but of surpassing excellence", and he added that he had never ridden a horse that pleased him more than Governatore. Other drafts of Mantuan horses were sent as further additions to the King's stud and usually English hounds and hobbies were sent in exchange. The Duke of Urbino also presented a stallion, which Cardinal Wolsey, an enthusiastic horseman, used. Wolsey was energetic in improving the breed of horses in Britain at this period.

In 1515, the King of Spain, Ferdinand of Aragon, sent two excellent horses caparisoned *regio ornatu.*

In 1519 Henry VIII sent Sir Gregory de Cassalis to Bologna to buy the best horses procurable in Spain and in Italy. He went to the Duke of Ferrara and was shown the Duke's stud. It seems that the horses were not up to the standard required but two were chosen: "of the breed of Isabella, Duchess of Milan". Now the Duke wrote to Henry, apologizing for the lack of selection, "but the breed of horses in this country is very much degenerated"—in only five years!

At Eltham there were in 1519 coursers, young horses, hunting geldings, hobbies, Barbary horses, stallions, geldings and two stalking horses, in all eighty-six animals. So the chronicle continues, "Paris arrived with horses [probably from Fontainebleau] with a present of six cortalles [Spanish horses], tow Turks, a barbary, tow genettes, a sturring horse [courser] and tow little mules".

In 1526, Henry made a present of eighteen horses to Francis I and thus we see that breeding at Hampton had been successful. In June 1530, the Privy Purse paid £57.17*s*.4*d*. for bringing three horses, two men and one boy from 'Mantwaye' and Hore adds that drafts of this strain were sent to Spain. And in 1539, the King received twenty-five beautiful Spanish horses from Charles V.

Elizabeth I became a great patron of the Turf. Her racing establishment was at Greenwich and it was here that the Barbary horses were kept. Forty racehorses were stabled with two jockeys. She improved the royal studs, probably entered and ran her own horses, and certainly attended race meetings. There were race-courses at Salisbury, Doncaster, Huntingdon, Croydon, Richmond (Yorkshire), Carlisle and Chester. "In the reign of Queen Elizabeth racing was carried on to such an extent as to injure the fortunes of many individuals." Matches were run for silver bells and one is still preserved in Carlisle. Horse-racing was popular in Scotland from about 1552 onwards.

As one turns the pages of the annals, one cannot help feeling that breeding horses and racing formed the chief amusement of town and country folk and it makes that oft repeated statement that imported Eastern horses of the eighteenth century *founded* the line of Thoroughbred horse breeding look rather silly!

However, in 1599 Jervis or Jervase Markham, who followed Thomas Blundeville as a writer of things equestrian, recommends the Courser of Arabia as the ideal stallion to breed from for the Turf, the Arab in particular, of whom he writes: "whom if you view in the Summe, you would judge him either like changeable Satin or Cloth of Gold". This horse was travelled from Angelica[23] to Constantinople, across Germany by land and by sea to England. A good horse, indeed, who would surely impart stamina into his offspring.

Markham discusses training and feeding horses for a course, "which we commonly call running Horses" and contrasts steeplechasers with racers. And then bread fed to horses cost one penny for a "goode bigge white loaf"!

Several Thoroughbred horses of Eastern blood were acquired from time to time by James I. The most notable, and one of the earliest mentioned in this reign, was the Markham (but whether it was the same Arab as the one owned by Jervis, is not clear) which the King purchased for £154.

The Duke of Newcastle says of this horse: "He was a *Bay*, but a *Little* horse, and no *Rarity* for *Shape*; for I have seen *Many English* horses farr *Finer* . . . and being trained up for a *Course*, when he came to *Run* every horse *Beat* him."

James also received from Naples "a dozen gallant mares, all with foal, four horses and eleven stallions, all coursers of Naples". These went to Greenwich whilst in 1617, half a dozen Barbary horses arrived in England and were sent to the royal paddocks at Newmarket.

Queen Anne received from her father-in-law, the King of Denmark,

[23] Said to be in Arabia, but it has not been located.

twelve mares which went to Byfield Park. Thus within a few years both Neapolitan and horses similar to those which formed the basis for the later Fredericksborg stud found their way into the royal stables where racehorses were bred.

In 1623, thirty to thirty-five horses from Spain were sent as presents to the Prince of Wales, Henry, who died early and was the elder brother of Charles I. But during his short life he was an enthusiastic horseman and breeder of racehorses. At Ickles in Yorkshire the Reresby family had a considerable stud.

During the reign of Charles I, English horses were much valued in France. Marshal de Bassonpierre tells us in 1626 that the court at Fontainebleau customarily played for large sums and the courtiers called the counters *Quinterots* because they passed from one player to another with as much celerity and rapidity as the *English horses were known to run*: they were called *Quinterots* from the name of the person who had brought them into France the year before.

On the execution of Charles I, the royal stud at Tutbury fell into the hands of Cromwell's men and a list or inventory was made of the horses there.[24] There were 23 mares with foals, 15 mares of four and over, 16 three-year-old fillies and colts, 17 two-year-old fillies and colts, 22 yearlings fillies and colts, and 23 horses four years old and over—altogether 139 horses. There are five horses with the name Morocco, and Hore considers that many were descended from the Digby and Villiers Arabs and some of Sir John Fenwick's famous Arabians. One is called Sorrell Fennick. One would have assumed that the Morocco horses might have been Berbers and the Sorrell Fennick too. Although I have seen a number of unusually marked Arabs, I do not recall having seen a sorrel (roan) unless chestnut is meant here, whereas sorrel is a fairly usual colour amongst Berber horses.

In the sixteenth- and seventeenth-century chronicles the recurrence of the word *barb* for a swift horse, e.g. racehorse, does not always appear to signify that the horse referred to was a Berber or Barbary horse. The idea that suggests itself to me is that because Barbary horses were renowned for their speed and endurance, the word *barb* was commonly applied to a fast horse. There is nothing faster than the barb of an arrow!

At this stage it is almost (although not always) impossible to distinguish between the strains of Berber and Arab horses. It seems probable that either nomenclature was used indiscriminately, North Africa (where in any case, nomad Bedouin bred pure and bastard Arabian horses) and Asia imposing no geographical difference.

The term *Arab* is not so frequently used and it really seems that the few genuine Arabs that appear in these chronicles were of great beauty

[24] J. P. Hore, *History of Newmarket*, Vol. II.

and stamina but were certainly not swift. Most Eastern horses of speed appear to be indiscriminately called Barbs—just as a century or two later most Eastern horses and North African as well of *any breed or strain, and there were dozens*, came to be called indiscriminately Arab.

The royal Tutbury stud was disposed of by the Parliamentarians in 1651 and some horses were acquired by Sir Arthur Haselrigg for his Woodstock stud. The stud or farm at which racehorses were kept was styled 'race': Tutbury Race, Cole Park Race, Malmsbury Race.

As we know, the Duke of Newcastle was exiled for some years at Antwerp, during which time he employed himself with horsemastership. He maintained that there were "Good and Bad of all Nations; but that the *Barbes* were the Gentlemen of Horse-kind and *Spanish* Horses the Princes". He bought, he says, "four Barbes, four Spanish Horses and many Dutch Horses, all the most beautiful that could be; and among them a Grey Leaping-Horse, the most beautiful that I ever saw; and who went exceeding High and Just in Leaps, without any Help at all".

The Duke puts the Spanish horse before all other breeds. He says they are best for manage, for war, for ambling, for hunting and for racing. And he gives us the names of the following famous running horses, all of whom were Spanish: Conqueror, Shotten, Herring, Butler and Peacock who was a mare. He adds that they "Beatt all the Horses in their Time, so much, as No Horse ever ran near them".

A Spanish stallion called Il Bravo, sent to the Archduke Leopold was considered to be worth a "Manor of a Thousand Crowns a Year". Newcastle helped Montagu to buy Barb horses in Paris, and these he writes of in a disparaging manner although one of them "did win Many Matches". He thinks they were not "Right Barbs, neither by their Shape, nor Price, but Bred in some islands thereabouts". The only praise he has for them is as racehorses!

The Duke of Newcastle had little experience of the Arab, saying the only one he knew was the one bought by James I from Markham, which he considered a poor specimen and when trained was unsuccessful on the turf. Here J. P. Hore notes: "It is impossible to distinguish the characteristic peculiarities of the so-called Barbs and Arabian horses in vogue at this time. There is no precise authority on the question. Doubtless they were all classed as 'Spanish' and of Eastern blood."

Newcastle goes on to say that the Hungarian, Polish, Swedish and Turkish horses are "the Most Beautiful in the World" but for him the English horse "is the best Horse in the whole World for all Uses whatever from the Cart to the Manage; and some are as Beautiful Horses as can be had anywhere, for they are bred out of all the Horses of all Nations".

Under the Parliamentarians in 1649–59, not only was breeding nearly ruined but horse-racing was forbidden, since it was supposed to conceal the reason for Royalist meetings. Although Cromwell himself had an eye for a running horse—he bred racehorses and imported Arabian horses[25] and Place's White Turk was one of his stallions at stud. Place was his stud groom. Cromwell also had a famous brood mare, called the Coffin-Mare, since she was concealed in a vault during the search for his effects at the time of the Restoration.

Earlier I referred to the early races as being run in matches. One horse was matched against the other for very high stakes and it is interesting to note that the weights varied from 8 to 10 stones, the distances ranging from 4 to 7 miles. For instance: Bull against Woodcock on 13 October 1681, weight 9 stones, distance 6 miles; and on the 15th or thereabouts in the second match Bull versus Woodcock, weight 9 stones, distance 7 miles.

In 1687, horses were run in three heats, carrying 12 stones.

On 17th December 1684, Evelyn went to St James's Park to see three Arabian horses which had just arrived from Hamburg, having been three weeks on the journey. He writes:

> I never beheld so delicate a creature as one of them was, of somewhat of a bright bay, two white feet, a blaze; such a head, eyes, ears, neck, breast, belly, haunches, legs pasterns and feet, in all regards beautiful and proportioned to admiration; spirited, proud, nimble . . . They trotted like does, as if they did not feel the ground. 500 guineas was demanded for the first, 300 for the second and 200 for the third.

This event was the forerunner of later importations which, out of several hundred, include Byerley Turk in 1689, Darley Arabian in 1705, and Godolphin Arabian or Barb in 1728.

Now we have already seen that the designation *Arabian* apparently provided an umbrella for many Eastern horses, without any relation either to the strain or the actual country where it was bred. Also the Ottoman Empire under the Turks covered the ports from which Eastern horses could be shipped, and just as Arabia was never part of Persia, so Syria apparently was never part of Arabia! It has confused the issue when later writers refer specifically to Arab horses when they really mean horses from Persia, Turkey or Syria. Earlier chapters have dealt with genuine Persian and Syrian breeds.

A typically mistaken description may be established in the paintings by John Wootton (1677–1765) of Lord Oxford's Dun Barb and the same noble Lord's Bloody Shouldered Arabian. *Both* horses belong to the same strain! W. Shaw Sparrow shows that a Turkey merchant, Nathaniel Harley, exported the Bloody Shouldered Arabian from Aleppo (Syria) in 1719. Nathaniel Harley wrote to his nephew Lord Harley, later Lord Oxford, to whom the horse was sent, as follows:

[25] Some were imported from Aleppo!

> By Capt. Oliver I send a Grey Horse that I have had by me more than four years, and I have two or three times attempted to send away, but have been disappointed and I think I wrote you of him before. He is of the Gordeen breed, of which there's few remaining, and is the chief cause of my sending him. He is of great spirit, but no great speed, wou'd soon learn anything in the Manage . . . I bought him of a Bey of the Gordeens when he was only two and a half years old, and because you will observe a Redish stain on one of his shoulders, which is now much wore out, but when I bought him was as red as blood.

Harley then explains that the owner of the mare that was to be the dam of the Bloody Shouldered Arabian was a robber and being wounded he leant over the mare's neck and his blood ran down her shoulder—thus her foal was born with the mark! However romantic this may sound, the red stain is known and called *navajo* amongst breeders of the American Spanish Mustang. *Navajo* means 'razor cut', a light background with red splashes!

The most interesting point emerges that Gordeen or Gordyene was an ancient part of Parthia from 500 B.C. and is shown geographically as Corduène under Trajan (A.D. 98–117)—nearly 2,000 years ago! The district lies to the south-west of Lake Van Gölü, on the River Tigris, and in Harley's day came under the Ottoman Empire. Yet the horse who sired the famous Sweepstakes and Brisk is classified as Arabian!

Sweepstakes sired O'Kelly's Old Tartar Mare "whose offspring, after she was twenty years old, brought 30,000 guineas to her owner".

The Dun Barb's name is equally misleading, since he was shipped from Aleppo in 1715. (The Barb or Berber supposedly originated in North Africa!)

Another painting by Wootton of the Darley Arabian again shows a horse of conformation similar to the two already mentioned. Although Wootton's racehorses *may* not have been faithful portraits, yet the similarity of conformation cannot be dismissed. He was said to be a pure-bred Keheilan of the Munghi strain and in that case was a runner; another source states that he was brought out of the Palmyra desert, a little to the south of Aleppo, from whence he was sent to England.

On the other hand the famous Godolphin Arabian was presumably a Berber from Morocco. From his conformation (in a number of paintings) he certainly was of an entirely different breed to the Grosvenor Arabian. There are a number of tales as to how and where he was found, but the fact appears to be that he eventually went to Lord Godolphin's stables around 1730 in the lowly role as teaser. The mare Roxana was to be mated to Hobgoblin whom she refused, and the teaser was brought out. The result was Cade who founded the Matchem line. Cade was said to be the only horse capable of galloping against the "flying wonder"—Flying Childers, a son of the Darley Arabian.

Possibly the Byerley Turk is one of the few horses to be correctly named. An English source, prejudiced in favour of *Arab* pedigrees, calls him the Arab charger of Captain Robert Byerley, who rode him at the Battle of the Boyne in 1690. A Continental source, however, states that the horse was captured from the Turks during the Siege of Vienna in 1689, by a Dutch officer who sold him for 100 gulden to Captain Byerley.

In *The Horsebreeder's Handbook*, second edition, there are listed ninety Arabian, forty-six Berber, thirty-two Turk, and four Persian sires amongst what Osborne is pleased to call Eastern sires. Thus Osborne, who is not alone in his opinion, says:

> There can be little doubt, despite all that has been and still is written upon the subject of 'Arab' blood in our English Thoroughbred, that at the best very vague notions exist as to the distinctions of breed in the Oriental sources . . . The distinguishing names Arab, Barb and Turk are found each in sufficient proportions as to nullify any special or singular claims.

From the eighteenth century on, there appear on the pages of Thoroughbred history great racehorses, celebrated sires and dams—and the people who helped to create them:

Eclipse "first, the rest nowhere", Herod, The Tetrach and Persimmon, whose son Perfectionist out of Perfect Dream founded the famous stallion line in the East Prussian (Trakehner) warm-blood breed of horses. Persimmon belonged to King Edward VII, and the horse's head and hoof are in the collection in the Queen's Gallery at Buckingham Palace.

There was the great Hungarian-bred mare Kincsem who won all fifty-four races in which she was entered, including the Goodwood Cup and many European classics. There are some charming stories about Kincsem, for she was a personality. Her blood runs in the "W" family of the Röttgen stud, Germany.

The mare Sceptre, bred by the Duke of Westminster in 1899, a daughter of Persimmon, won the Two Thousand, the One Thousand, the Oaks, St Leger, Jockey Club Stakes and Champion Stakes and was beaten only in the Eclipse Stakes by Ard Patrick.

Signoretta, who was bred in Italy in 1905.

Hurry On in 1913 and Man o' War in 1917, both the most important horses of their generation in their respective countries, Great Britain and the United States of America. The latter won the Hopeful Stakes, Belmont Futurity, Preakness, Belmont Stakes, Lawrence Realization and the Kenilworth Park Gold Cup. He was bred by August Belmont.

Appelle was bred by Frederico Tesio in 1923; her breeder needs no introduction. The mare raced twenty-three times and won thirteen races, including all the most important races in Italy and the Coronation Cup.

Ortello, bred in 1926 by Nob. G. de Montel, who amongst other races won the Arc de Triomphe. Hyperion by Gainsborough. Brantôme bred by Baroness E. de Rothschild in France. Nereide bred by the Erlenhof stud in Germany.

Donatello, Italy, who also carried Tesio's colours, as did Niccolo del Arca, and Boticelli, winner of the St Leger and Ascot Gold Cup amongst fourteen victories. Big Game, also a Classic winner and sire of winners.

In 1952, the great Ribot was foaled; he had sixteen wins out of sixteen starts and his influence at stud is proved by his sons Ribero, Riboco, Ribocane and Ribofilio; and Molvedo bred in Italy. Ireland bred Tulyar, Ballymoss and Charlottesville; Citation and Nashua were bred in the States. Tantième and his son Match were bred by François Dupré, one of France's foremost breeders; also Tanerko and his son Relko were bred in France.

Coming to 1968, there is Sir Ivor whose name is linked to that of his jockey Lester Piggott. Sir Ivor's American owner, Mr Raymond Guest, affirmed that the horse would run in both continents that year. Royal Palace won the 1967 Derby and the Coronation Cup at Epsom ridden by A. Barclay who won the Hardwicke Stakes with H.M. the Queen's Hopeful Venture.

Among the great steeplechasers are Golden Miller, Brown Jack, Arkle and the many winners, mares and geldings—and a few entires as well who won the Grand National at Liverpool, too many to name individually!

We come to Vaguely Noble, bought at Newmarket for 136,000 guineas, and since his win in the Prix de l'Arc de Triomphe said to be worth 2,000,000 dollars. Vaguely Noble was bred by the late Major Lionel Holiday, trained by Étienne Pollet in France; he is owned by Dr Franklyn and Mr Bunker Hunt of America.

In the language of hippology we refer constantly to 'blood'—the introduction of Arab blood; Thoroughbred blood; Irish Draught blood. What we really mean is that we, the breeders, are introducing the *genes* of this or that other breed into another breed. A breeder who is a genius at his work will 'nick' the genes so successfully that the product will, in the first or succeeding generations, attain the desired result.

I have never, in the course of many years of research, come across a breeder of horses who was successful on a 'hit and run' basis. A 'try it and see' principle can only once in a thousand times achieve the desired result. Also an irreparable amount of damage can be done and has been done in the past. Breeds have been destroyed, not only with horses but with cattle too (and this we are at present in the process of witnessing).

Although science has provided us with the knowledge of the blood groups of horses, no work whatever has been done on blood typing within breeds of horses (or cattle). We do know from experience that certain breeds possess certain genes in dominant or recessive forms which perhaps other breeds may not possess. The convex profile, for example, trotting action, jumping action.

Perhaps the following statement verges on the rash or the bold, but the gene producing the convex profile *appears* also to produce a sloping croup. If coupled with strong hocks, this, in turn, can produce high action. A high actioned horse is invariably a good jumper—a good jumper, whether steeplechaser or show-jumper, must have the physical quality of strong hocks and joints whatever else he lacks. In the same way we know that the gene producing strong, flat bone also produces a *hardy* horse. It is immaterial if the horse in question is a middleweight hunter or a Western Isles pony.

One of the fallacies of the modern age is the one where the breeder puts a big horse to a small mare with the idea that the small mare will then produce a bigger foal than herself. A big stallion does not necessarily carry a gene for size whereas a big mare does, and the small mare carries not only the gene for smallness but she has neither the room nor the milk to cope with the production of a large frame and body. This has been proved many times. Strangely, of all the unwise matings that breeders have indulged, an Arab × Shire cross has never been used on a commercial basis. Unconsciously, then, we do take genes into account. But equally unsuitable crosses have been and are frequently perpetuated. The eighth Duke of Beaufort wrote, about ninety years ago, that a well-bred mare and a common stallion would breed faster progeny than could be expected from a Thoroughbred stallion and a common mare.

A very interesting experiment which I saw in Poland, and one which must have given tremendous satisfaction to the breeders themselves, was as follows. Some well-bred East Prussian mares around 16 h.h. (whose ancestors originally had Tarpan blood) were found in an appalling condition on the Don steppes after the war, having been taken as booty. They had spent several post-war winters out in below-zero conditions and sometimes with little or no grazing. These mares were returned to Polish studs in their East Prussian homeland but neither an East Prussian (Trakehner) nor a Thoroughbred stallion was successful in getting them in-foal. As a last resort, a Tarpan stallion of 13·2 h.h. was turned with them and was entirely successful. The resulting progeny were substantial horses from 14·2 to 16 h.h., generally light or blue dun in colour (Tarpan), whilst one mare, a big mouse dun with a markedly Tarpan head and wither stripe, had been mated to a Thoroughbred. I saw the foal at two months—an upstanding *blood* colt (illustration 109).

Thus the potent primitive genes of the Tarpan had nicked, one imagines, with dormant Tarpan genes of the original East Prussian mares,

108. A Cleveland Bay mare, whose dam was by the first Harpham Turk. Painted by Dally of York *c.* 1790

109. The mare is by a Tarpan stallion out of an East Prussian mare, her foal is by a Thoroughbred. (Taken in Masuria, Poland)

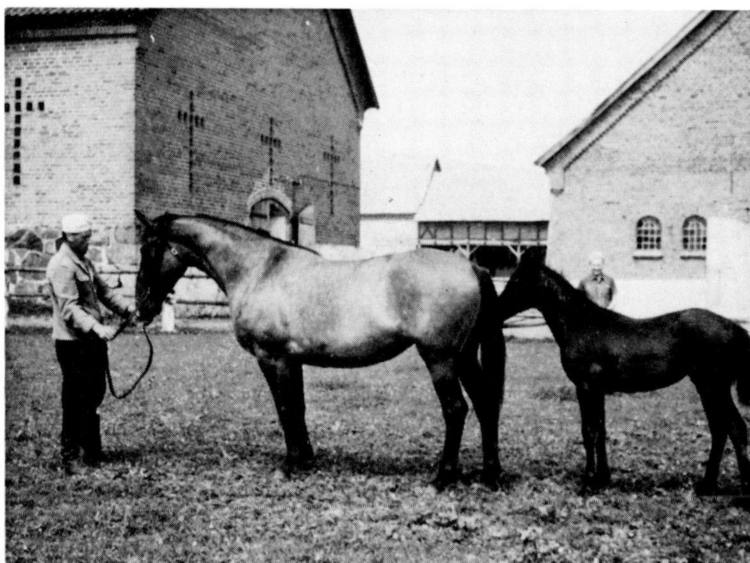

110. A modern brood mare: Tuppenny II, by Toddhurst (Mossborough) out of an Irish mare Penny, seen here with foal at six weeks by Grand Roi (Charlottesville); photographed 1972

111. Old Billy, a race-horse, said to have been the oldest horse recorded. He died at 83 years of age. Painting by Charles Towne

112. *(below)* The famous Irish-bred Thoroughbred steeplechaser Arkle, who belonged to Ann, Duchess of Westminster

who themselves also carried Thoroughbred genes through Persimmon[26] and thus when a vigorous cross was mated again to a Thoroughbred, the result was 100 per cent successful. What has to be bred into horses is first soundness, second soundness, and third soundness and *no reason is good enough* to breed from unsound stock.

The two illustrations on pages 232 and 233 are reproduced by the kindness of Dr W. E. Jones and Professor R. Bogart from *Genetics of the Horse*. Both J. Hammond and J. E. Flade made similar studies and both obtained the same kind of results: Shetland stallion × Shire mare, Shire stallion × Shetland mare. The progeny of the Shetland dam grew to resemble the Shetland breed more closely and at thirty-eight months weighed 724 pounds, whilst the progeny of the Shire dam grew to resemble the Shire breed and at thirty-eight months weighed 1,008 pounds.

The above-quoted examples should show us that maternal influences should be a most important consideration to every horse breeder. The point is also further highlighted by the fact that a hinny more nearly resembles its dam the donkey and a mule more nearly resembles the mare than it does its sire, the jack.

Nature herself is a ruthless selector and horses that will not breed probably cannot breed successfully; it is a breeder's job to assist and encourage nature not to overrule her. Unhappily no breeder has to have qualifications, no mare a certificate of soundness, and horses are expected to conform and produce the greatest effort of exertion no matter what their inheritance of (un)soundness, conformation and ability.

[26] See *Flight of the East Prussian Horses* by Daphne Machin Goodall.

Conclusion

In conclusion we may ask: "What is a breed?" We have the species: *Equus*. The sub-species: horse, ass, onager, zebra. The strain or breed within the sub-species with its differentiating and peculiar characteristics. M. Wilckens wrote: "If conditions essential for life, under the influence of which the breed form occurred are consistent, then the type is fixed and there is a continuity of inheritance. As soon as these essential conditions change so also will the form and performance of the breed change." Oppositely the Settegast definition adds to the last sentence that especial emphasis is laid on the adaptability of horses to certain given conditions.

But Wilckens says further: "The durability or permanence of a breed is dependent only on certain conditions for life to which it is adapted: *there is no such thing as absolute permanence of any form of breed* [author's italics]."

The Wilckens definition of a breed is regarded in scientific circles as of the most significance since it agrees with the Mendelian point of view.[1] Pure-bred refers to and means only individuals of the same breed; cross-bred with a member of a different breed; half-bred implies a Thoroughbred cross with another breed; and in-bred where members of the same family are bred together. This practice may have disadvantages as well as the advantageous emphasis on genes to fix a desired characteristic.

Adametz adds that "it is not always recognized the tremendous role that good hay and good grass play in rearing horses".

It is not always understood that there are certain grasses which horses enjoy eating and do well on and other grasses which do them no good at all. The grasses upon which cattle flourish are not the same as those for horses. The old idea that horses should follow cattle since they can graze closer is a good one *if* care has been taken to sow the right mixtures. And for horses a varied mixture which includes *Poa pratensis* is suggested, as well as clovers. Horses must be fed good hay with a really aromatic 'nose'. New hay should not be fed for at least three months after making, and 'new' oats not until after Christmas. If oats are overfed, an indulgence in excess protein may cause the bones to be affected. In fact, it is

[1] A point of the utmost significance is that it also applies to the human species.

erroneous to suppose that massive feeding of oats will create either substance or energy. It will do neither. For no horse can absorb and digest more than its quota of protein, which must be accompanied by the correct amount of added minerals and vitamins if the desired result is to be obtained. Injections to put on bloom or weight, to 'force' yearlings for the sale ring or for show, are ill advised since there is no proof that these could or would not affect subsequent training and performance and possibly even the ability to breed successfully.

It is dangerous to play with nature but it is common sense to assist nature to improve her product, as long as we know what we are doing and hope to achieve.

As we have seen throughout the pages of this book, nature has been prodigal in the evolution of *Equus* and the subsequent gift to mankind can scarcely be exaggerated. For adaptability, endurance, speed, strength, collaboration, generosity and kindness the horse has proved to have been man's greatest ally throughout their journey together along the paths of history. Man himself evolved from nothing; if he destroys nature and the bonds which nature forges, he will return to nothing.

It was not my intention, nor indeed would it have been possible, to give details of the breeding or even to mention by name many breeds of horses or ponies. Almost *all* the breeds in the world are mentioned in *Horses of the World*, together with the appropriate photographs. The intention in this book is to be concerned with the evolution and subsequent history of certain strains as these crossed or developed into breeds which have had a distinct genetic influence on modern *Equus*. Some breeds of horses have had a very short history and almost no traceable influence; with others, the star of hereditary influence has burned brightly for many centuries and that influence is none the less valid if it has been divorced from legend. Other hidden genetic values have also perhaps been brought to light.

For these reasons some specified modern breeds have not been included. They may indeed possess a short history and in time they may have an improving influence on the species—this will be up to the breeders—but at present, useful though the breed in question may be, it does not yet qualify for a place in history.

In the past, much has been written and repeated about the development of horses which later discoveries have shown not to be true. Earlier writers did not have access to this knowledge so they can be excused; but modern writers cannot be excused for giving opinion as fact, nor for failing to give information as truthfully as possible when that information exists: *ignoratio neminem excusat*.

Sometimes I have made what may appear to be contrary statements and it is because absolute certainty on the point in question cannot be established. I have therefore given *both* sides, so that perhaps time and further research may clarify the position.

It is clear that in more than three years of extensive research which covered thousands of years of history I have still not been able to produce all the information which must exist in some language somewhere about the history of horse breeding in a particular locality. I have, however, been at considerable pains to check opinions and facts and to keep the twain apart. An unruly opinion and an untrue fact make bad stable companions.

If this book does no more than awake an interest in breeding horses in the ordinary layman, then it will have achieved something, but if it gives genuine breeders some knowledge and appreciation of the genetic history of the particular breed which they themselves are furthering, then the work will have been really worthwhile.

And I repeat once more: "*There is no such thing as absolute permanence of any form of breed.*"

Appendix 1. *The Chronological Period when Horse Breeding Began*

3000 B.C.	Dnjestr—Dnjepr area: Tripolje A, B1, B2, C1 & C2 down to the Scythian occupation of this area.
500	North Europe wild horses
2500	Dümmersee to
2000	Bronze Age
500	British Isles: late Neolithic wild horses and early Bronze Age to Iron Age
2000 +	Jordansmühl, Danube area
1500 +	Hallstatt civilization
2000 +	Caucasus wild horses: Kuban, Maikop earliest breeding down to Scythian period
500	Upper Volga
2000 +	Ural Mountains: River Ob and Altai-Sajan-Jenissei to
1500 +	Scythian and Sarmatan period
2000 +	China: Yang-Shao wild horses
1500 +	Lung-shan. Shang Dynasty 1523–1300 to
1500	Chou Dynasty 1027–246 B.C.
5000	Euphrates–Tigris area: Onager breeding at Qualat Jarmo and
3000 +	Turan. In the main onager but horse breeding centres at
2500 +	Kisch, Chafadshi
2000 to	Babylonia, Kültepe, Tschagar Bazar
1500	Charsamma

Appendix 2. *Relative Classification of Names between China and Asia*

Horses: Ta-yüan	=	Ferghana, best horses found in Ir-shi (Nish: Nisaea)
Nisaen	=	bred on the Plains of Nisaea, very well-bred horses up to weight.
People: Wu-Sun	=	Ferghana. Lived near Lake Issyk-kul (T'ien Shan mountains). Bred the Ili horse. They came from Kua-chóu to the east of Lob Nor in the fifth to fourth centuries B.C.
K'ang-ku	=	Sogdiana
Hu;* Hung-nu;* Huns*	=	Tartars
Ta-Yüé-chi	=	(Indo) Scythians, Epthalites or Hepthalites
Ta-hia	=	Bactria
K'iang	=	Tanguts
An-si†	=	Parthia

This knowledge has become available only during the past hundred years and then generally only in Russian and German sources. This may account for certain misinformed statements such as John Lawrence (*Treatise on Horses*, 1810): "Arabia is the oldest breeding country in the world." There is absolutely no foundation whatever that horses were to be found either in the wild state or domesticated in early historic times in Arabia.

There are further unmistakable indications as to the original areas of horse breeding. There are five definitions of words for horse.
Akwa, ak, Indo-European language group: '*to hurry*' which developed to

* All of Turkish stock, were different phases of the same tribes.
† ANŠE = horse? Babylonia
Ansi = a station on the Silk Road

Sources: *Journal of the American Oriental Soc.*, Vol. 137, 1917.
Prof. Fred Hirth, *The Story of Chang K'ien.*

ekwos, horse, spanning those countries from India to Ireland, from Iran, the Mediterranean to the far north.
Morin, horse, Mongolian reaches from Japan and China across the Kalmuck steppes to Russia.

The last three belong to the Japhetic group such as the Sumerian, Churrite and the language of other mountain peoples of the ancient Orient.
Rob-moni is a Slavonic expression as is *koni*, old Russian: *komoni*. (Old Prussian: *camnet*: and *kumelys* for foal (*kumyss*: milk), but there is a basic form *kobmuji*, *kobmonji*, and in old Anglo-Saxon *kobyla*: mare. J. Schmidt supposes that *kobmoni* was an ancient pre-indo-germanen description. *Koni* may in the East represent the Chinese *kuendi*: mountain horse. In the East and South-Europe the tap root word is *kob*, *cabo*. Personally I feel that this is where the expression *cob* came from—a short-legged mountain horse.)
Mandus belongs to the Etruscans and Mediterranean peoples.
Kaval belongs to Armenia and Central Asia and also Slavonic and Finnish language group.

"To be precise the science of language of the Avesta points to the homeland of horses: 'there, where the longest summer day equals the two shortest winter days'—which happens just below 50°N in the Euro-Asian forest steppe belt," (Hančar).

An ancient legend of the Scythians, who were once the inhabitants of High and Central Asia, says that

> in the Creation of this part of the earth's surface, when these high lands divided from the flooded waters, it was here that human life first appeared. In this manner so all subsequent migrations of peoples have gone out from this part of the world. For these very reasons, one may put the centre of creation here, it is probable that horses as well as others of our domestic animals had their original habitat, and that it is from this area that they have slowly spread to other parts of the world. (Ammon.)

There is much to be said for this ancient legend. Zimmermann, the most famous naturalist scientist of his time, agreed with it and it cannot be exactly a coincidence that here in High Asia, the high plateau of the Gobi Desert, and nowhere else in the world, the primitive Asiatic wild horse may, we hope, still be found. This is the tap-root of all breeds of horses, and as well it is in this area that most of our other domestic animals are to be found.

Bibliography

Early Period

F. Flor, *Haustiere u. Hirtenkulturen*, 1930, Wiener Beiträge zur Kult. u. Linguistik.

Max Hilzheimer, *Unser Wissen von der Entwicklung der Haustierwelt Mitteleuropas*, 1927.

Ernst Weidner, *Weisse Pferde im alten Orient*, 1952.

Joseph Wiesner, *Der alte Orient*, 1939, J. Hinrichs Verlag, Leipzig.

P. Yetts, *The Horse: A Factor in Early Chinese History*, Eurasia Sept. Ant. Helsinki, 1934.

Helen. E. Fernald, *Chinese Art and the Wu-Sun Horse.*

W. Koppers, *Pferdeöpfer u. Pferdekult der Indo-Germanen*, 1936, Pustetver., Salzburg.

R. Lydekker, *Guide to the Specimens of the Horse Family*, 1907.

—*A Geographical History of Mammals*, 1896, B.M. Nat. His., 190.

—*Natural History of Quadrupeds.*

Bengt Lundholm, *Abstammung u. Domestikation des Hauspferdes*, Zoo. Ant. Fran Uppsala 27.

E. Schwarz, *Über diluviale Pferde des Equus Caballus Gruppe*, Jahrbuch des preuss. geo. Landesanstalt 48, 1937.

R. Chapman Andrews, *Natural History of Central Asia.*

R. C. Andrews, *Essays on the Cenozoi of N. China*, 1923, Mem. Geo. Survey of China. S.A. No. 3.

C. P. Berkley & F. K. Norris, *Geology of Mongolia*, 1927, Nat. His. of Cent. Asia. Amer. Mus. of Nat. History.

E. W. Pfizenmeyer, *Mamutleichen u. Urwaldmenschen*, 1926, Leipzig.

V. Christian, *Archiv. für orient. Forschung* 10, 1934.

M. Gryaznov, *The Ancient Civilization of South Siberia*, Barrie & Rockcliff, London.

F. W. König, *Älteste Geschichte der Meden und Perser*, 1934, AO 33. H. 34.

Herodotus, *Travels.*

F. Schachermeyr, *Poseidon u. die Entstehung des griechischen Götterglaubens*, 1950, Munich.

E. Erkes, *Das Pferd im alten China*, 1940.

W. Watson, *China*, 1961, Thames & Hudson.

C. F. Hawkes, *The Prehistoric Foundations of Europe.*
V. G. Childe, *Prehistory of Scotland.*
V. G. Childe, *Horses, Chariots and Battleaxes,* Ant. 15, 1941.
Franz Hančar, *Das Pferd in Prähistorischer und Früher Historischer Zeit*, Verlag Herold-Wien München, Bdl, 1955.
L. Adametz, *Lehrbuch der allegemeinen Tierzucht*, Wien, 1926.
A. Alföldi, *Antike Darstellung zur Kenntnis der Kultur der euroasischen Reiterhirten*, Fol. Arch. III & IV.
J. W. Amschler, *The oldest pedigree chart, Journ. of Heredity* xxvi, Washington, 1935.
O. Antonius, *Die Rassengliederung der quartaren Wildpferdes Europas*, Verh. d. K.K. Zool. Bot. Ges. Wien. lxii. 1912.
— *Zur Abstammung des Hauspferdes*, 1936, Ztschr. f. Zucht u. Zuchtungsbio. etc xxxiv. Berlin.
— *Wagenkämpfer und Wagenkrieger bei Indogermanen und Germanen*, Odal, Goslar.
A. Nehring, *Studien zur Indogermanischen Kultur und Urheimat,* WBKL lv, 1936.
H. G. Creel, *The Birth of China*, London, 1936.
Boule, Breuil, Licent & Teilhard de Chardin, *Le Paléolithique de la Chine*, 1928.
A. Kammenhuber, *Hippologia Hethitica,* 1961.
A. Salonen, *Hippologica Accadica,* Helsinki, 1956.
Breuil, Berger, Kirchner, Holm, Lhote & Lommel, *Steinzeit*, Holle.
Eggers, Will & Joffroy, Holmquist, *Kelten und Germanen*, Holle.
K. Jettmar, *Die Frühen Steppenvölken*, Holle.
C. J. du Ry, *Völker des alten Orient*, Holle.
W. Torbrügge, *Europaische Vorzeit*, Holle.
E. Porada, *Alt-Iran*, Holle.
W. Keller, *Und die Bibel hat doch Recht*, 1966, Rowohlt.
O. R. Gurney, *The Hittites*, Pelican.
J. Wiesner, *Die Thraker*, 1963, Kohlhammer, Stuttgart.
—*L'Art Préhistorique.*

Middle Period
A. Hink, *Über schwere Pferde im alt- und mittel-persischen Reich*, 1909, Zeitsch. f. Gestütskunde und Pferdezucht.
K. W. Ammon, *Pferdezucht der Araber u.d. arabischen Pferden Documenta Hippologica*, 1834, New ed. Olms Presse.
M. Jankovich, *Pferde, Reiter, Völkerstürme*, BLV.
Xenophon, *The Art of Horsemanship*, J. A. Allen.
Claudius Aelianus, *On the Nature of Animals.*
E. Hirth, *The Story of Chang K'ien*, 1917, vol. 37, Journal of the Amer. Orient. Soc.

A. Weigall, *Wanderings in Roman Britain.*
F. Runger, *Herkunft der Ritterpferde des deutsche Ordens*, 1925.
F. Harper, *Extinct and Vanishing Mammals.*
T. Blundeville, *The Fower Chiefest Offyces of Horsemanshippe.*
Marcus Fugger, *Von der Gestüterey*, 1527.
Newcastle, Duke of, *Méthode et Invention de Dresser les Chevaux.* (A New Method and Extraordinary Invention to Dress Horses and Work Them according to Nature) 1667.
B. D. de Castillo, *Historia Verdadera de la Conquista de la Nueva España.*
J. Marek and H. Knizkova, *Tschingis Chan und sein Reich*, 1963, Artia Praha.
P. Albert Huber, *1000 Jahre Pferdezucht Kloster Einsiedeln*, EDE Verlag.
E. Foord, *The Last Age of Roman Britain*, Harrap.
J. Burckhardt, *The Age of Constantine the Great*, Routledge & Kegan Paul.
Alföldi, *Zur his. Bestimmung der Avarenfunde*, Eu. Septent Antigua 9–10, 1934.

Modern Period
M. A. Denham, *North of England Folklore*, 1858.
Ilsan Abidin, *Die Pferdezucht u. Pferderassen im osmanischen Reiche*, 1918.
Dr Fuller, *History of the Worthies of England*, 1662.
von Hutten-Czapski, Count, *Geschichte des Pferdes.*
Capt. M. Hayes, *Among Horses in Russia*, 1900.
Dr G. G. Simpson, *Horses.*
I. T. Sanderson, *The Dynasty of Abu.*
H. Trunz, *Pferde im Lande de Bernsteins*, Paul Parey, 1967.
R. d'Andrade, *Alrededor del Caballo Español*, 1954.
— *Garranos.*
— *O Cavalo do Sorraia.*
— *O Cavalo Andaluz de Perfil Convexo*, 1941.
Roczniki Nauk Rolniczych, Polish Ag. An. 85 & 108, 1963.
W. Pruski, S. Schuch and I. Grabowski, *Hodowla Koni*, Bd 2.
W. Pruski, *Hodowla Koni*, Bd 1.
Count Wrangel, *Die Rassen des Pferdes*, Bd 1 & 2, Schickhardt & Ebner, 1908–9.
D. Carruthers, *Unknown Mongolia*, 1914.
E. H. Parker, *A Thousand Years of Tartar Rule*, 1969, Dawsons.
Jaques de Solleysel, *Le Parfait Maréchal*, 1664.
Watson and More, *The Science and Practice of British Farming*, 1942.
Jos. Osborne, *The Horsebreeder's Handbook*, 1889–90.

Lord Ribbesdale, *The Queen's Hounds.*
S. Wright, *Evolution in Mendelian Populations*, 1931.
W. E. Jones and R. Bogart, *Genetics of the Horse*, Caballus, 1971.
Le Cheval de Selle Français de Normandie (Assoc. des Eleveurs du Cheval Normand).
F. Kaudewitz, *Grundlagen der Vererbungslehre*, 1957.
A. Wrangel, *The Arabians in Arabia*, J. A. Allen.
E. Schiele, *Araber in Europa*, BLV.
H. Aubanel, *Die weissen Pferde der Camargue*, AMV, 1963.
Sir R. Glyn, *The World's Finest Horses and Ponies*, Harrap.
E. Canti, *Nel Mondo del Trotto*, SISAR.
— *I Grandi Campioni del Gallopo*, SISAR.
Dr E. Mohr, *The Asiatic Wild Horse*, J. A. Allen.
The Genius of China, Catalogue to Chinese Ex., British Museum, 1973.
J. Patton, *The Light Horse Breeds.*
L. Shinkarew, *The Land Beyond the Mountains.*
J. P. Hore, *History of Newmàrket and Annals of the Turf*, 1885, Vols 1 & 2.
Wm. C. Miller, *Practical Essentials in the Care and Management of Horses on Thoroughbred Studs*, 1965.
F. C. de Beaulieu, *Der Klassische Sport*, 1942.
— *Vollblut*, 1967, BLV.
F. Tesio, *Breeding the Racehorse*, 1958.
Mally, *Pferdezucht.*
W. Zorn, *Pferdezucht*, 1952, Eugen Ulmer, Stuttgart.
H. Neuschutz, *Pferdezucht*, 1956.
F. Haines, *Appaloosa*, Amon Carter Mus. 1963.
T. Ryder, *The High Stepper*, 1961.
— *King of Sports, U.S.A.*
Col. H. Cousté, *Mechanics Applied to the Racehorse*, 1916.
Pers Crowell, *Cavalcade of American Horses.*
Vivien Kendrick, *Horses in Japan*, J. A. Allen, 1964.
Daphne Child, *Saga of the South African Horse*, Howard Timmins, 1967.
M. J. Barones van Heemstra, *Het Friese Paard.*
H. Epstein, *Domestic Animals of China*, 1969, C.A.B.
F. Traut, *Gestüte Europas*, 1971, Deutsche Pferdemuseum. Hipp. Ins.
S. Lehndorff Graf, *Ein Leben mit Pferden*, 1956, Landbuch Ver. Hannover.
R. Black, *Horseracing in England*, Richard Bentley & Son, 1893.
J. Flade, *Das Araberpferde.*
Moses Griffith, *The Welsh Cob*, S.B. Vol 38, 1956.
T. Halford, *Montgomeryshire Horses: The retrospect*, 1900.
Sir R. D. Green Price, *Horses for the Country: How to breed them*, 1903.

Dobie, Bodright and Ranson, *Mustangs and Cowhorses.*
Robert M. Denhardt, *The Horse of the Americas*, University of Oklahoma Press.
Ensminger & Uppenborn, *Handbuch der Tierzuchtung*, Paul Parey, Bd 3.
A. Herrmann, *An Historical Atlas of China with text*, 1966, Djambatan NV.
Grösser historischer Weltatlas, BLV. Bd. 1.
J. C. Morton, *A Cyclopaedia of Agriculture*, Vol. 2, 1855.
Larousse, *Encyclopaedia of Prehistoric and Ancient Art.*
Shepphard, *Historical Atlas.*
Reader's Digest, *Great World Atlas.*
Time/Life, *Nature Library Series.*
The British Racehorse.
Sankt Georg.

Index

A

B

C

D

E

F

G

H

N

Q

R

S

T

U

V

W

X

Y

Z